Gates's

JIGS, FIXTURES, TOOLS & GAUGES

Sixth Edition Revised and Enlarged

by

G. H. RYDER

Principal Lecturer in Applied Mechanics,
Royal Military College of Science,
Shrivenham

TECHNICAL PRESS
LONDON

First Published 1925
Second Edition 1939
Reprinted 1941
Third Edition Enlarged 1944
Fourth Edition Revised 1949
Fifth Edition Revised 1959
Reprinted 1964
Sixth Edition Revised, Enlarged and Re-titled 1973

ISBN 291.39432.9

Text set in 11/12 pt. Monotype Times New Roman, printed by photolithography, and bound in Great Britain at The Pitman Press, Bath

Preface

For the first time since its original publication in 1925, Mr. Philip Gates' successful little book *Jigs, Tools and Fixtures* has been comprehensively rewritten and re-arranged for this, its Sixth Edition; and even its title has been somewhat altered and amplified in an attempt to convey the full range of the subjects covered and the order in which they are discussed in the book. Its new editor gratefully acknowledges the free hand which he has been allowed by Mr. Gates (now well on into his nineties but still keenly interested in the health and continued development of his brain-child) in his efforts to make both text and illustrations fully suitable for present-day use.

The object of the book remains the same as it always has been from its first edition onwards—to explain and illustrate the basic principles underlying the design and operation of certain essential component parts of, or adjuncts to, the modern machine tool, in all its great variety of types and purposes.

My aim as its new editor has been to offer to the apprentice in the machine-shop and to the junior in the drawing office a general introduction to the main types of equipment used in connection with machine tools; to pass on certain useful advice picked up in the course of a long and varied career in the shops and as production engineer, author and lecturer; and to underline the more important points by appropriate illustrated examples.

I believe that the book also contains much that will be of value to people working on the clerical side of the production process (progress chasers and the like); while more senior men on the shop floor itself—the experienced machinist and even the skilled tool setter—may derive from some of the more advanced sections the stimulus of new information or of a new method of approach.

Despite the enormous progress which has been made since 1925 in the design and variety of the machine tools and their related equipment which one would expect to find installed in a modern plant designed for the rapid, accurate and economical production of an article of which large numbers of identical copies are required, the basic principles of jig and tool design remain largely unaltered; and those parts of the book devoted to a discussion of general principles have merely been reworded in this latest edition. Additions to the scope of the book, however, have been continuous throughout its life; and the principal changes made to this Sixth Edition are the inclusion of up-to-date information on the performance of modern cutting tools and on selection of the correct type of tool for a given job; details of the latest practice in the manufacture and use of gauges; and an entirely new chapter describing the organisation of the production process from the management point of view and discussing the influence of jig and tool design on manufacturing costs.

In conclusion, I would like to echo with complete agreement the following paragraph which Mr. Gates wrote towards the end of the Preface to the last edition of his book, for it strikes with great accuracy the note on which I would like to conclude my own introductory remarks. "Although," wrote Mr. Gates, "the author is well assured that none of the information herein contained is consciously inaccurate, nor any of the suggested solutions to specific problems other than practical and sound, he would desire to remind his younger readers that there exists a not inconsiderable diversity of opinion on many points of real importance in this fascinating and swiftly-growing field. It follows therefore, that the accepted solution to many of these problems is by no means necessarily the only solution; and it is hoped that many of the methods recommended, and opinions expressed, in the following pages will be read with this reservation in mind."

G. H. R.

Shrivenham
May 1972

Table of Contents

CHAPTER 1

The Principles of Production Tooling Construction

1.1 Definitions. The purpose of the four types of equipment whose design and application are considered in this book is to speed production in manufacturing industry and to ensure that accuracy and quality are achieved at economic cost.

Since it is necessary to be entirely clear from the outset about the distinction between jigs, fixtures, tools and gauges, their characteristics and fields of application are summarised below:

(*a*) A **jig** is a device which holds a component during a machining operation, and which provides guides for the cutting tools such that a desired feature can be accurately machined into the correct position on a large number of successive components. The most common application of a jig is in multiple drilling and boring; and since it is often necessary to machine several faces of the workpiece, it is not usual to clamp a jig to the machine table.

(*b*) A **fixture** is a component-holding device which is clamped to the machine table. Though the component is held in the fixture, there are no tool guides. Separate means must therefore be provided to set the tools in relation to the fixture, and some additional skill will be needed to carry this out. Fixtures are used extensively in milling operations, and to a lesser extent in turning, grinding and broaching.

(*c*) A **tool** is a device which is brought into contact with the work and is there used to form the shape and size of component required. Only machine-operated tools are considered in this book; and these fall into two main groups: **press tools** and **cutting tools.** Press tools are used for forming components out of sheet metal by such processes as blanking, drawing and bending. Cutting tools are used in machining operations to produce finished components by removal of material from bar stock, rough castings or forgings. They may be classified either according to the process in which they are used (*e.g.* as turning, milling or drilling tools) or merely as "single-point" or "multi-point" cutting tools.

(*d*) **Gauges** are used to check the dimension or the shape of a particular feature on a component or assembly, either by direct measurement or by comparison with a precision gauge. They are of especial use in repetitive work such as the inspection of quantity production, or in tool-setting applications where they are much quicker and more convenient to use than any manual checking method. When a toleranced dimension is to be checked, a matched pair of gauges is used, with one of the pair set at the maximum size and the other at the minimum. The matched pair is then known as a **limit gauge. Profile gauges** (usually cut from a plate) are used for checking the shape of a component.

1.2 Modes of Production. The type of production undertaken in any given workshop may be defined according to either the level of output or the variety of work undertaken. The characteristics of each type determine the necessary level of capital investment in machines, the use of special equipment and tooling, and the class of labour required.

(*a*) **Unit or Job Production** is the term used when jobs are mostly "one-off" or made in small numbers, with repeat order unlikely to be obtainable. Almost all tool-room work, for instance, falls in this category. The machines used need to be of standard type, and as far as possible suitable for multi-purpose use. Jobs are normally specially marked-off for machining, and are located in standard equipment. Jigs, fixtures and special tools are only rarely used when essential to the satisfactory completion of the job. Labour must be highly skilled and versatile, with an ability to work under minimum supervision and technical guidance.

(*b*) **Batch Production** is the most common mode of production used in industry. Machines are set up to make a special number (called a *batch*) of a particular component and are then reset to make a batch of a different component, and so on. Batches may either be made when required to meet a specific customer's order, or they may be made at regular intervals for stock or as part of a larger assembly. The size of batches can always be adjusted to meet demand, or to minimise cost of manufacture and storage.

High-production machines such as automatic lathes are widely used to produce large batches of components machined from bar stock. For smaller batches, plug-board-controlled and numerically-controlled machines are coming increasingly into use.

Jigs, fixtures and handling devices are used to reduce machining and handling time, and to ensure constant quality of production. This is especially important in the later stages of assembly, where interchangeability of components is essential. Their use also enables semi-skilled labour to be employed.

(*c*) **Flow** *or* **Continuous Production** is a method of working used only in the vehicle and domestic product field. An entire factory is laid out so that components and sub-assemblies move by set routes from the raw material stage to the finished product. A high volume of continuous production is achieved, leading to the use of transfer machines and elaborate jigs, fixtures and mechanical handling equipment. Little skilled labour is needed, most of the work being of the machine-minding or routine assembly type.

(*d*) In **Mass Production** very large quantities of standard components such as nuts, bolts, washers, etc. are made on special single-purpose machines. A high degree of automation is desirable, both in manufacturing and handling.

1.3 Design Principles for Jigs and Fixtures. The most important considerations are *accuracy* and *rigidity*, followed by *ease of use* and *economy in construction*.

Loading and unloading procedure should be quick and easily accomplished, especially when the machining cycle is short. Methods of clamping should be safe and foolproof, so that danger to the operator and the likelihood of work fit only for scrap being produced are both minimised. Measures must also be taken to avoid distortion or damage to the component.

The weight of a jig or fixture should always be kept to a minimum consistent with the necessary rigidity, in order to facilitate handling and storage. Methods of removing the metal shavings (known as *swarf*) produced during a stock-removal operation of any kind, and of cleaning after use, must both be carefully thought out.

Jigs and fixtures are usually made sufficiently robust for detailed stress analysis to be unnecessary; but sharp corners and sudden changes of cross-section should nevertheless be avoided in design, for both lead to concentrations of stress which may result in failure by fracture.

The choice of material to be used is discussed in detail in Chapter 3, but two points of principle should be mentioned now. First, all surfaces which are particularly subject to wear, such as locating and guiding surfaces, must be made of specially hardened material. Second, the danger of distortion caused by changes of temperature must always be borne in mind, particularly when dissimilar materials are used in the composition of an individual jig or fixture, or when the component is made of brass, for example, or aluminium.

Lastly, the capital cost of a jig or fixture must always be considered in relation to its use and expected life. When not in use it has to be stored, which involves handling and storage costs. Some deterioration must be expected to occur from wear, corrosion and accidental damage; and allowance should be made for replacement of the more vulnerable parts if the life is to be prolonged.

1.4 Design Features of Jigs and Fixtures. Five considerations are important, of which the first two are common to both jigs and fixtures, the third applies to jigs only, and the last two only to fixtures.

(1) **Location of the Component.** The component must always be located accurately in relation to the jig or fixture—a consideration which means that any relative movement between the two must be eliminated. Since any body in three dimensions has six degrees of freedom (that is to say, it can move in any of three directions and can rotate about any of three axes) location can only be achieved by fixing *six points of contact*, as will be seen in Fig. 1.1.

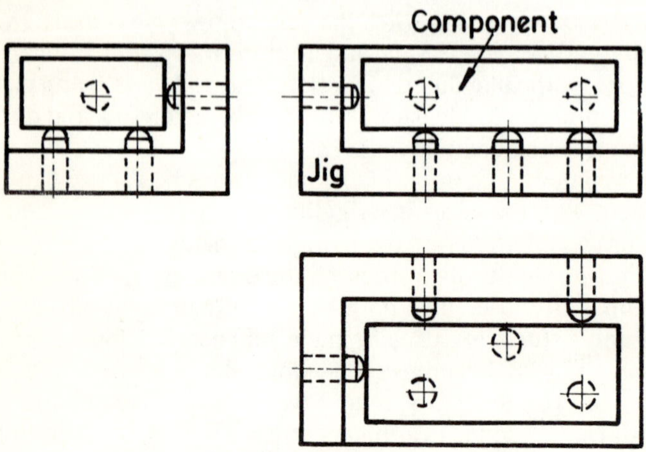

Fig. 1.1 The Principle of Six-Point Location

The form of location shown is suitable for an approximately rectangular component, but other forms are not difficult to devise for other shapes. It is important to ensure that there are sufficient points of location yet at the same time no unnecessary or redundant constraint, for the latter can result in distortion or imperfect seating—

particularly when components vary in size as would be the case when rough castings are being machined.

Two other simple and frequently used forms of location are shown in Fig. 1.2—(*a*) for flat lever-shaped components and (*b*) for hollow cylindrical components.

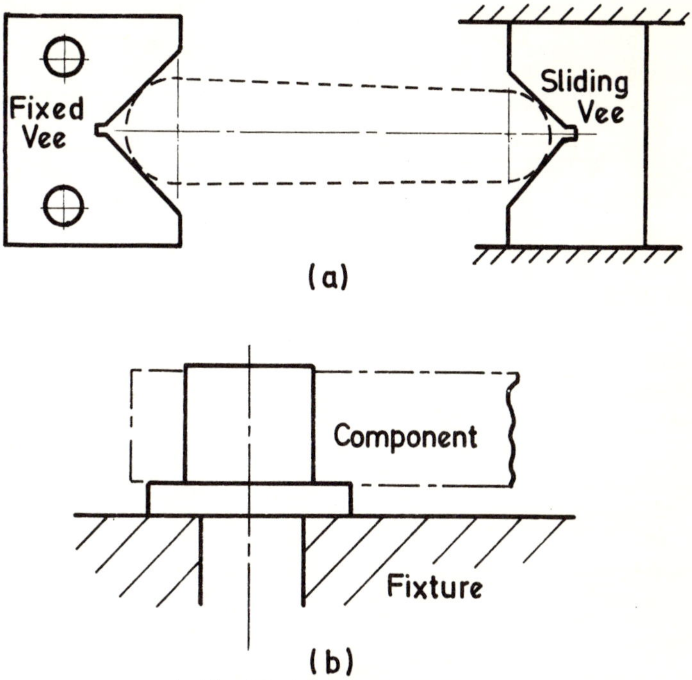

Fig. 1.2 Methods of Location
(a) By Vee Blocks
(b) On a Plug

(*b*) **Clamping of Component.** Location alone is not sufficient, since it usually depends on gravity or friction to maintain contact with the locating surfaces. A clamping force has also to be applied, and must exert sufficient force to resist the cutting action without causing distortion either to the component or to the jig or fixture.

Some simple types of clamp are shown in Fig. 1.3. Others will be discussed in Chapter 3 and later as details of larger assemblies for specific applications.

(*c*) **Guidance for Tools.** This requirement for drilling jigs is met by inserting hardened steel bushes into the body of the jig. Chapter 3, again, will provide further information.

(*d*) **Setting of Cutters.** In the case of a fixture, the tools are not guided into the work. It is therefore necessary to be able to set the cutters in relation to the fixture. Setting gauges are used for the purpose, in conjunction with accurately located surfaces on the fixture itself.

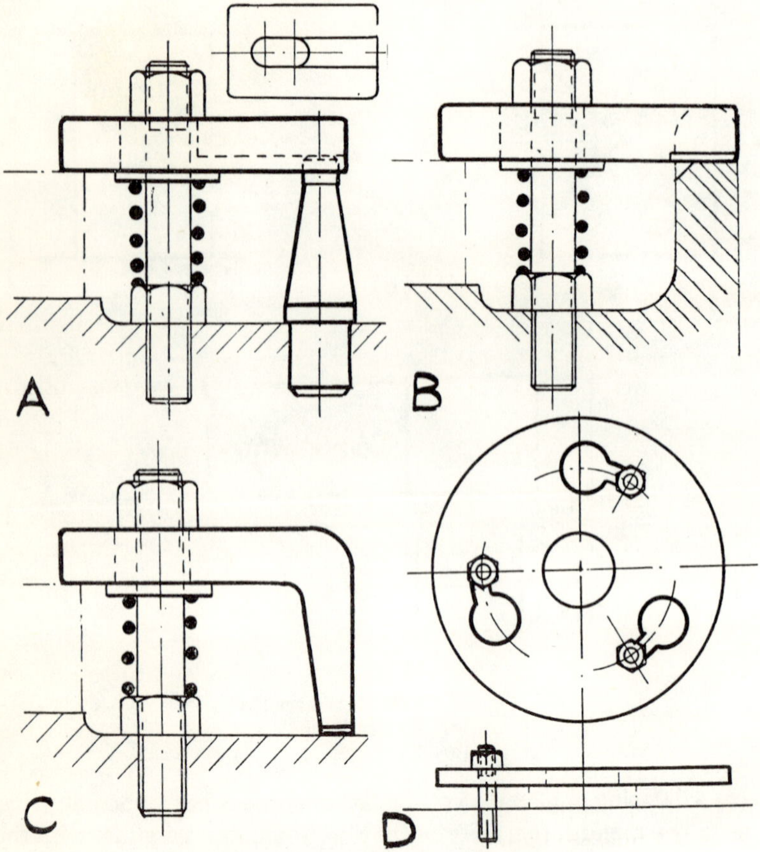

Fig. 1.3 Some Simple Types of Clamp

(*e*) **Securing to the Machine Table.** Means must always be provided to bolt or clamp a fixture to the machine table.

CHAPTER 2

Drawing and Sketching

2.1 Introduction. The information given in this Chapter is intended to serve only as an introduction before the reader proceeds to more specialised literature. He is advised in the first place to make himself familiar with the recommendations made in BS 308 issued by the British Standards Institution, on which this chapter is based.

2.2 Drawing Equipment. Different people will always get different results even with the same grade of pencil, but for general work a good 2H or 3H will be found satisfactory. For outlines or bold work, and for lettering, an H is useful.

The correct sharpening of a pencil is a great help to neat and accurate work. The accepted shape is a long wedge, with the flat side of the lead being used against the rule edge to produce lines of unvarying thickness. To help maintain the point, a flat small file is a useful possession; but there are also available small glass-paper pads from which layers can be stripped as they become worn.

There is a wide choice of scales. Printed celluloid should be avoided, because the marking wears off after a week's contact with the paper and there is constant need to check the straightness of the edge. For taking measurements of machines, etc., a flexible steel rule and a slide gauge or callipers will be found useful.

Practice with regard to drawing papers varies in different offices, but three grades can be said to be in general use: *Cartridge paper*, *Detail paper*, and *Tracing paper*. The former is generally used where the drawing is intended to be permanent, or when it is thought likely to be subject to many alterations necessitating much erasure. Detail paper is a finer and cheaper grade much used for small drawings or for roughing-out ideas generally, although extra-good quality detail paper is sometimes used for permanent drawings from which prints can be taken if the lines are made suitably bold. Tracing paper is useful when a portion of a permanent drawing on which modifications need to be made is taken out. Rough ideas can also be sketched on the tracing paper.

As for drawing pins, always choose those having a head of large diameter, for it is the head that holds the paper, not the pin. Alternatively, drafting tape can be used to hold drawings firmly in position. It has the advantage of avoiding damage to the drawing board.

It is always best to buy drawing instruments from a maker of repute. After using them with ink, take care to clean them thoroughly. Buy first only the instruments you *must* have, then add to them as experience develops. In this way the cost is spread over a period of time and you should be able to build up a full kit of quality instruments without undue financial stress.

2.3 Selecting Views. Since the essential point of a mechanical drawing is that it shall convey as clearly as possible the ideas of the draughtsman to the mechanic, the selection of the correct view is obviously important. The drawing of an unnecessary view is not only wasteful but misleading, whereas inadequate or insufficient views always give rise to queries. The correct method of showing plans and elevations is the subject of a difference of opinion between what is known as *third angle* and what is known as *first angle*. The distinction between the two is illustrated.

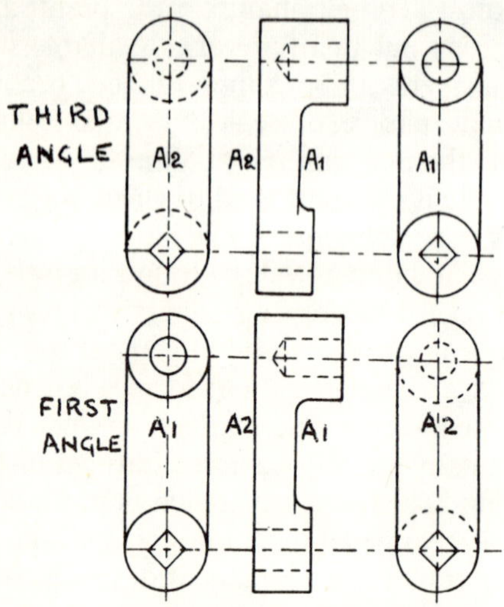

Fig. 2.1 First and Third Angle Projection

In the upper group in the illustration (**third-angle projection**), when the component shown in the centre of the group is viewed from the right, its face A1 will be seen and is pictured *on the right of the group*. In **first-angle projection,** on the other hand (see lower group in Fig. 2.1), the view of the component from the right (still labelled face A1) is shown *on the left of the group*, while the view of the component from the left is pictured on the right of the group.

The existing practice in a given office will obviously be the first consideration; but subject to that it matters not which arrangement of views is given as long as what is intended is made perfectly clear.

In cases where the usual plans and elevations are not sufficient, section or part section views can be added, as in Fig. 2.2.

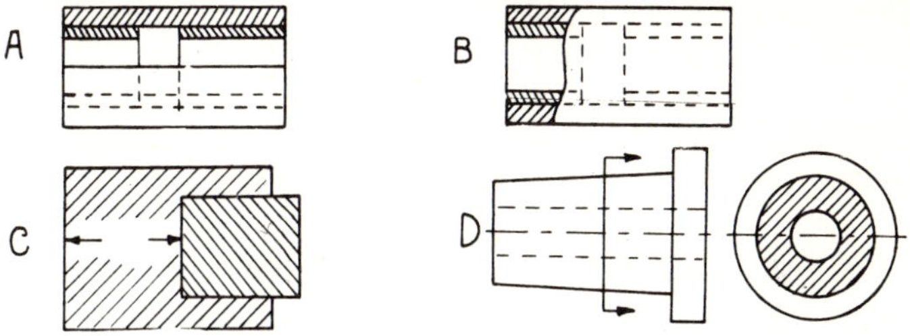

Fig. 2.2 *A*. **Half Section.** *B*. **Part Section.** *C*. **Dimension in a Section.** *D*. **View Point of a Cross Section**

2.4 The Completed Drawing. Fig. 2.3 is a drawing of a cast-iron holder suitable for a certain size of turret lathe. It embodies all the

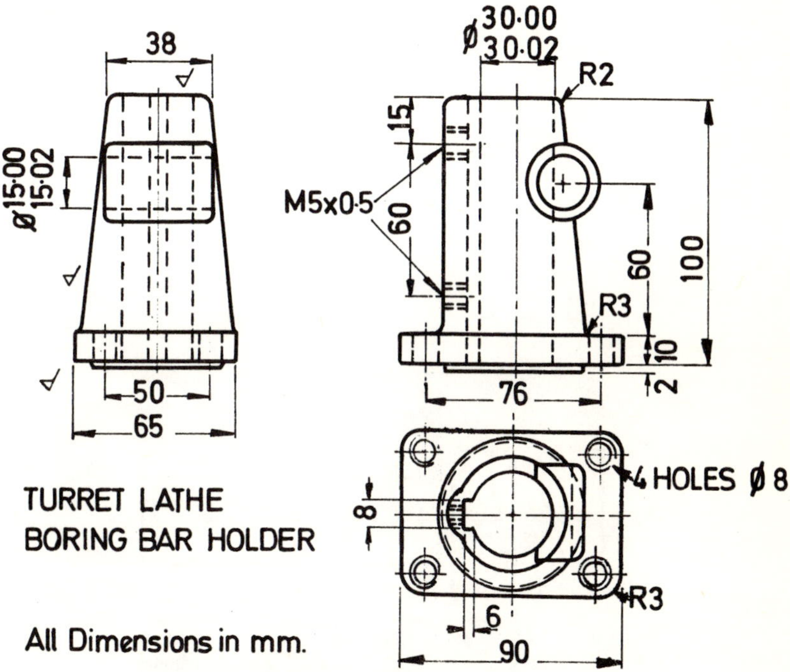

Fig. 2.3 Example of a Drawing done to BS 308

recommendations made by the British Standards Institution save that the title block in which the draughtsman prints the title of the drawing, his firm's name, the Drawing Number, etc., is omitted.

The principles of mechanical drawing hold good whether it be engine parts or jigs and tools which are being drawn; but in the latter case, a different technique is often possible. When the size of the component for which equipment is to be designed permits, it is a good plan to draw one or more views of the component to scale in red ink, and then as it were to "erect" the jig or fixture round it.

2.5 Lines. For indicating the shape or profile of an object, a bold line such as that shown at **A** in Fig. 2.4 should be used. This prevents the general shape from becoming lost in the maze of other essential lines. The line shown at **B** is used for dimensioning and the projection of views; while for hidden outlines the line **C**, consisting of regular short dashes, is used. It should not be confused with the short chain line **E** which would be used to indicate an alternative position of a unit, such as the position of a crank at different positions in its path.

The chain line shown at **D** is especially important, for it indicates the centre of bores, holes and important parts. It is customary to insert these at an early stage in executing a drawing, and then to build round them.

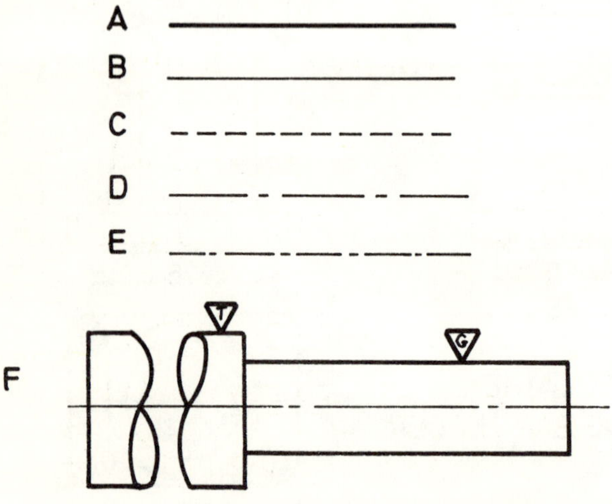

Fig. 2.4 *A* to *E*. Types of Line used in Engineering Drawings. *F*. Method of showing "Break," and Machining Indicators

2.6 Machining Indicators. The variety and grades of finishes which are available for surfaces make necessary something more helpful than the old-fashioned "f". One method is shown at **F** in Fig. 2.4, in which the finish required is indicated by the inserted reference letter **T** (for Turn). Had the letter been **G** (for Grind) it would have been accompanied by an indication of the surface finish required.

2.7 Dimensioning. A number of important principles are involved when a decision is taken on which dimensions are to be shown on a component drawing, and how these dimensions are to be toleranced so that the required fit and location is obtained on assembly—even though the style to be used is laid down in B.S. 308. These principles and conventions are illustrated in Fig. 2.3. They may be summarised as follows:

(*a*) Sufficient dimensions must be given to determine completely the size and shape of the finished part, but *only* sufficient dimensions. Redundant ones can lead to incorrect manufacture and misinterpretation of tolerances. A possible exception arises in a case where all intermediate dimensions are given; but in this case it is recommended that the overall length, put in for information only, be clearly marked as an "auxiliary" dimension.

(*b*) Always select an important functional surface as a datum, and dimension other functional surfaces from it. This ensures correct location on assembly.

(*c*) The dimensioning of non-functional surfaces is not critical, and can be carried out in a manner best suited for manufacture or inspection.

(*d*) Always specify the tolerances on dimensions which affect functioning or assembly. It is sometimes convenient to state on the drawing a general level of tolerance to be applied, with any exceptions given against particular dimensions.

(*e*) Where features such as holes and pins have positional tolerance as well as size tolerance, the positional tolerance should be followed by the letters MMC, which stand for Maximum Material Condition. If this is done, parts which are made to less than MMC sizes may be acceptable even though the positional tolerance may be slightly exceeded.

2.8 Sections. Sections are useful for showing internal construction and shape. At one time the type of line used for this purpose also indicated the class of material, but the types of material used nowadays have multiplied to such an extent that this method has been abandoned in favour of plain lines at 45°, spaced according to the area being covered.

Section lines can be interrupted to allow a dimension to be inserted, as shown at **C** in Fig. 2.2.

The half-sections **A**, and the part sections **B** in that Fig. are useful in that they can sometimes be used to avoid the necessity for drawing other views.

When a member is shown as being cut completely through, the direction from which the section is to be viewed should be indicated as at **D** in the same Fig.

2.9 Representation of Screw Threads and Gears. All drawing office charges are added to the cost of a given job, so unnecessary work must be avoided. It is therefore standard practice that such items as screw threads and gear teeth are not drawn in detail. Screw threads are shown in representational form at **A** and **B** in Fig. 2.5, and a threaded hole at **C** and **D**. Where a plain drilled hole is tapped, the depth of the drilled hole apart from the tapped portion should also be dimensioned, as at **C**. A piece of tube having an *external* thread is indicated as at **E**. A brief statement of thread size, form, etc., is neatly added, as near as possible to the thread or screw itself without causing confusion.

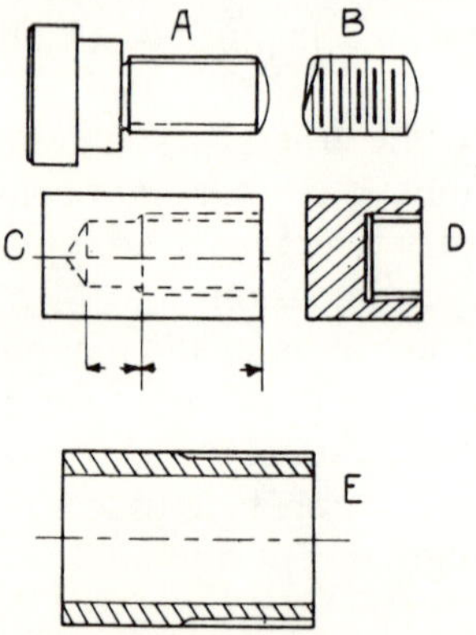

Fig. 2.5 *A–B*. **Standard Representations of Screw Threads.** *C–D*. **Standard Representations of a Threaded Hole.** *E*. **Standard Representation of an Externally-Threaded Tube**

For gear wheels, it is sufficient to draw plain circles representing the Outside Diameter and the Pitch Circle, and then to insert the necessary tooth information either in the circle or nearby.

2.10 Breaks. Whenever a unit or a view of a unit is disproportionately large, it is convenient to make use of a "break". Samples of these are shown at **F** in Fig. 2.4. To the left is the method of showing a broken solid bar, and to the right the method of indicating a broken tube or cylinder.

2.11 The Value of Sketching. The ability to make rapid and fairly accurate sketches is essential to the successful draughtsman, for whom sketching has been truly described as an auxiliary language.

It is often necessary to go over to the machine shops, not necessarily near at hand, to take details of a machine or tool. To do this the machine may have to be stopped, so the draughtsman must not only be quick but thorough; for if he forgets a point he may have to go back and request another stoppage—a request which will not normally be viewed with pleasure.

When discussing the pro's and con's of equipment, too, a few rapid sketches to make one's ideas immediately clear are of great value. Sketching is one of the finest ways of learning and remembering machine tool construction and details. Many people would be surprised to find how weak is their detailed knowledge of various constructions despite having looked at them, passed them by many times and even perhaps used them. Though quite certain that the mechanism is fully understood, they may well encounter difficulty if called upon to put the details down on paper.

Many are the arguments which have been settled by means of a few telling sketches with chalk on the machine bed or on the wall, or even on the palm of one's hand.

2.12 Proportion. It is always important that the various parts of a sketch should be in correct proportion to one another. While it might be said that if the dimensions are stated correctly almost any sketch is good enough, yet an ill-proportioned one can be very deceiving. A mechanic turning a job on the lathe or working on the shaper, for instance, soon gets into the habit of letting the machine slide travel until the piece begins to look something like what is shown in the drawing, and only then applying the rule or other measuring instrument. Fig. 2.6 shows at **A** an example of a sketch of the wrong proportions, with a comparatively short length marked 100 mm and a part looking considerably longer marked only 50 mm. The diameters in the sketch are also deceiving.

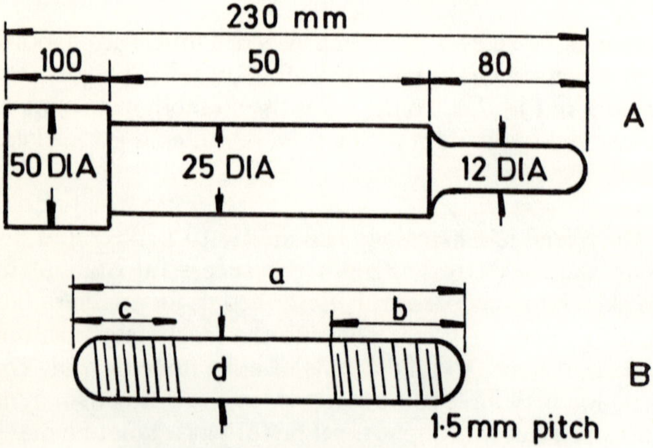

a	b	c	d	No. off
150	80	30	20	6
100	60	25	16	10
80	30	20	14	25

Fig. 2.6 *A*. **Example of a Badly-Proportioned Sketch.**
B. **Single Sketch used with Table of Dimensions**

Note in this sketch, by the way, that only one view is given. This is permissible where no error can possibly occur, as is the case in this example where the word or abbreviation for diameter is distinctly written after the figure.

Sketching pads made up of paper ruled in squares are useful, but a good draughtsman should practise so as to be able to sketch on any paper and in any circumstances.

2.13 Duplicates. Where rough sketches are called for in cases of urgency, the draughtsman will be well advised to put a sheet of carbon paper between the sheets of his pad so as to secure a duplicate. Such sketches often tend to become unreadable in their passage through the shops, and if a mistake occurs by reason of such hard usage, the value of the duplicate to the draughtsman is obvious.

2.14 Tabular Drawings. It is often necessary to call for a number of parts very similar in form, but differing perhaps in one or two dimensions. Much work can then be saved by making a single sketch of the part and adding to it key letters denoting certain of its dimensions, as shown at **B** in Fig. 2.6. A tabular statement of the sizes required is then inserted below or near the sketch.

2.15 Oblique Projection. The method of laying out an object to show views in a certain relation to one another, as in Fig. 2.3, is termed *orthographic projection*. While this is the usual method adopted, it is sometimes necessary, particularly when sketching an awkward piece, to make its general form obvious at once. For this purpose a method known as *oblique projection* is employed, as shown in Fig. 2.7. Note that it has the appearance of a perspective drawing save that most of the lines going in the same direction are parallel. This style of drawing makes it possible to explain a point to anyone unacquainted with mechanical drawing, and also enables three dimensions to be given in one view.

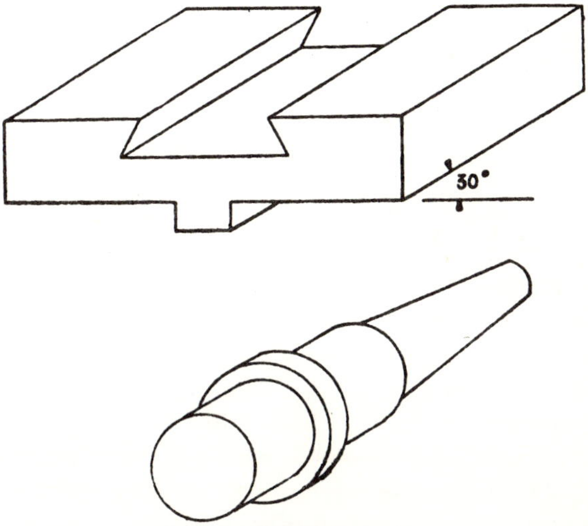

Fig. 2.7 Oblique Projection

In the rapid sketching of objects thus projected, the best results are obtained by placing the longest side of the piece horizontal. This prevents a distorted appearance. Then draw one of the edges upright, following with a side at 30° to the horizontal. If after this, all lines running in the same direction are made parallel, there will be no difficulty in producing quite lifelike sketches.

The best way of appreciating the value of oblique projection is to try to show by alternative methods a complicated piece of pipe-work having branches in different planes, or any even fairly elaborate piece of smith-work.

2.16 Drawing Office Practice. In developing an idea for a tool, jig or fixture, the designer must rely on past experience of similar work, and may well recall some design which has proved successful in practice and which, with modifications, may be re-used in different circumstances. By examining the office records, he may find that equipment already exists for a component of similar form which can be copied or modified to do both jobs. Remember, however, that while original ideas are welcome, anything in the nature of a freak design should always be avoided.

Once a clear idea has been gained of what is proposed, it is usual to make a layout or *General Arrangement*, one or two views of which will always disclose any snags which were not originally apparent.

2.17 Detailing. The General Arrangement of the fixture having been completed, and the idea having been approved, the drawing is then passed to the *detailer*, whose job it is to dissect it and make separate drawings of each part. To do this successfully, the detailer must thoroughly understand the nature of a mechanical drawing, and must see that no part escapes his attention.

The various parts of the jig or fixture are always numbered on the General Arrangement, and the detail drawings must bear corresponding numbers for purposes of reference.

Reference frames such as are shown in Fig. 2.8 are then added to the drawing and filled in. This may seem superfluous to the uninitiated; but a drawing is not only intended to convey the ideas of the designer to the mechanic. It should also bear sufficient details of its past history to enable easy reference to be made. Different offices have their own pattern frames, there being some difference of opinion as to exactly what and how much should be included on them.

The Component Number and Name to which the drawing refers is then added, as is also the Operation Number (see Fig. 2.9) and an indication of the machine on which this operation is to be carried out. Revisions and Modifications are entered bottom left. It is most important that these should be recorded, together with the date of revision.

The question of the different "fits" should be carefully standardised, for a lot of time can be wasted, both in the office and in the shops,

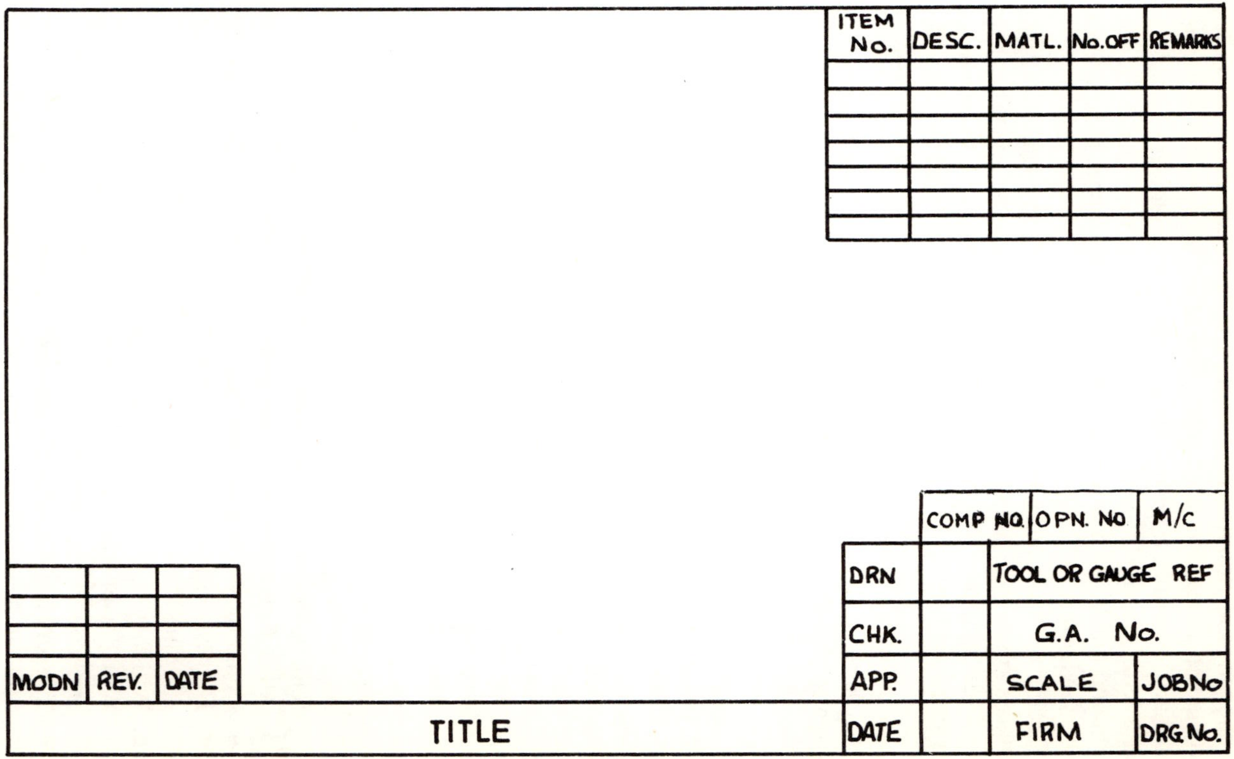

Fig. 2.8 A Typical Drawing Layout

by allowing individuals a free hand. A system such as that described in Chapter 11 can be used with advantage, which makes it only necessary to quote the decimal size of (say) the diameter of a hole, to add a letter denoting the class of fit, and then to indicate the tool room and machine shop working on the same system, to know just what is expected.

Details of the materials to be used and of any heat treatment required are of great importance in tool and jig work, and it has been found advantageous for all drawings giving such details to be marked in a standard way.

2.18 Checking. When detailing has been completed, the drawings are passed to the *checker*, whose job it is to examine them carefully for faults and errors of any kind.

Though a checker is not primarily engaged for the purpose of criticising, sensible comment on the efficiency of a design certainly forms part of his work. On receiving a drawing, then, the checker should also get hold of a drawing of the component which the tool or jig has been designed to make. It is better still if one of these components itself is available. Thus prepared, the checker will examine the drawing from the point of view of general construction. The design itself he will not interfere with, for he may assume that it will already have been approved by the head of his department. He will, however, work through all or most of the questions indicated below, though not necessarily in the order given:

1. In the case of a jig or a fixture, can loading and unloading be quickly carried out?
2. Are the locating spots arranged at the most important and reliable points?
3. Are the clamping arrangements satisfactory?
4. Could they distort the component when it is in position?
5. Do they allow the component to move while it is being machined?
6. Will they take too long to manipulate?
7. Will the operator have to think about them and so waste unnecessary machining time?
8. Can any loose pieces, if there are any, be eliminated?
9. Will the machining thrust be exerted against the clamps?
10. Is there adequate "chip capacity", or room for the escape of cuttings or *swarf*?

11. Is allowance made for any variation in the shape of the component?
12. Are all the details of the equipment given in the standard way?
13. In the case of a cutting tool, has the length of its life been considered when its cutting edges were being positioned?
14. Is the type of heat treatment called for the correct one?
15. Are the various fits generally suitable for their duties?
16. Have the recognised office methods been observed throughout?
17. Are the subsidiary dimensions correct when they are added up and compared with the overall dimension?
18. Are any unnecessary dimensions given?
19. Have all the details enumerated been included on the drawing?
20. Is the material called for correct?
21. Are the filing details correctly filled in?

Any error detected is pointed out to the draughtsman, who then corrects it. When the checker is at last satisfied, he adds his signature to the drawing and it is ready for the *tracing* department.

2.19 Tracing. Tracings, which are essential if clear prints are wanted, are normally done on tracing cloth, though temporary rough tracings are sometimes made on paper. The work is not difficult, requiring only common sense plus a pair of steady hands.

To make a tracing, the cloth is laid with its polished side facing the drawing, and is firmly pinned in position. The pins should be pressed well home so that the pinhead helps in the holding of the cloth. Failure to do this often results in the cloth tearing slightly at the pins, so altering the relation of the tracing to the drawing.

Next, starting at the top left-hand corner of the work, put in all the circular parts, using the ink compasses and bows for the purpose and working steadily towards the right-hand lower corner so as to prevent smearing. The tee and set squares are then employed to put in all the straight lines. This order of working is recommended because it is easier to make lines blend with curves rather than *vice versa.*

Titles and large printed letters can be put in with the aid of stencils, of which a great variety of different sizes and type faces exist. After a little practice in their use, neatness is easily attained. All dimensioning and small lettering, on the other hand, must be inserted by hand.

The tracer should never be tempted to put in lettering of his (or more usually her) own design, but should cultivate the habit of following the drawing in strict detail since the slightest addition or omission can cause serious trouble.

There is nowadays a variety of pens to choose from, but the important consideration is usually the diameter of the holder, because it influences the individual's ease of manipulation. The reservoir nib is of value when extensive areas of printing are unavoidable. Note, however, that with the modern fast-drying inks, frequent cleaning of the nib is essential.

If the surface of the cloth should appear greasy and the ink fails to flow, dust on some french chalk and then brush it off just before beginning work.

A blot, when it occurs, can be removed by immediately absorbing the ink with a bit of blotting paper and then wiping the spot itself with a damped cloth.

Never allow spots of water (*e.g.*, rain) to get on to a tracing; and never fold a tracing. If you do, both crease and spots will appear on the print.

2.20 Alterations and Revisions. No alteration or revision of a drawing or tracing must ever be issued for use save over the signature of a responsible person, and the fact and date of issue must be carefully recorded. The alteration is generally entered into a special book or index; and after the tracing has been corrected and the detail entered in the column under *Revisions and Modifications*, any unaltered prints still in circulation must be immediately recalled and replaced by the new one. It is possible for enormous losses to be caused by work being machined to old drawings never efficiently recalled.

Another essential precaution is that *everybody* whose work could possibly be affected by an alteration *must* be notified of the change, because the habit is easily formed of remembering dimensions by heart without reference to the drawing—a habit which, though convenient, is greatly to be deplored.

2.21 Filing and Recording. The importance of keeping correct records cannot be overestimated. The extent to which indexing is carried depends on the thoroughness of the management, some firms simply requiring a card saying that a certain tool or fixture has been made on such-and-such a date, while others demand details down to the number of the final storage bin in the tool stores.

The fact that waste of time and muddle can be avoided by an

efficiently planned and maintained filing system seems too obvious to stress—until you have worked in an office that lacks one. . . .

2.22 Operation Sheets. Printed sheets set out as a guide to the equipment available for the operations ordered on a given component, and the order in which they are to be used, are normally fixed to the component drawings issued to the shops. The way the sheets are made out depends on the "lay-out" of the particular operation; and when every tool required for it has been designed and made, its number is entered on the sheet.

When complete, the operation sheets are traced. Prints are taken of them, and they are then ready for issue. Fig. 2.9 illustrates a typical example.

OPERATION SHEET.							
OP. NO.	DESCRIPTION.	JIG NO.	TOOL NO.	FIXTURE NO.	GAGE NO.	M/C NO.	REMARKS.

Fig. 2.9 A Representative Operation Sheet

It is important that the order of operations should be entered correctly on an Operation Sheet, because grave trouble can result if (as is not unreasonable) relatively junior personnel follow blindly the order of operations laid down. If this happens with an incorrectly planned Operations Sheet, "progress-chasing" becomes almost impossible and the accuracy of the work itself is almost bound to suffer.

CHAPTER 3

Jig and Fixture Construction and Components

3.1 The material of which the main body of a **jig** or **fixture** is made will be either **cast iron** or **mild steel.**

The use of cast iron is largely confined to fixtures, whose weight presents no problem to the operator once they have been bolted to the machine table. Occasionally, jigs also will be made from castings if this offers economy in manufacture (which would only be likely if several jigs were to be made from the same pattern); but the permissible weight of a jig must always be limited by the need for ease of handling.

The main advantage of cast iron as a material is its high damping capacity, which enables it to absorb heavy and intermittent cutting forces such as may be applied to it when a milling fixture is being used. It also possesses good hard-wearing properties, and it needs no further treatment at locating surfaces. Its principal weaknesses are a tendency to brittleness, and low tensile strength. In the design of a casting, sharp corners and violent changes of cross-section should therefore be avoided. Rigidity must be built in by means of webs and stiffeners, and bending stresses should be kept to a minimum.

When mild steel is used as an alternative, the body of the jig or fixture is built up from parts made out of plate, bar, angle or channel sections; for mild steel is readily available in standard sizes, is easily machined and processed, and can be surface-hardened where required. Fabrication in mild steel usually gives a finished product of lighter weight than does cast iron, and it is for this reason in particular that it is preferred for jigs. In the case of fixtures, the choice will largely be decided on the basis of cost.

Construction and assembly in mild steel can be completed in one of two ways—either by bolting together the parts and ensuring accurate location by means of spigots and dowel pins, or by welding. Welding is now widely accepted as generally offering the cheapest and lightest assembly; but care must be taken during welding to avoid distortion and to give stress-relieving treatment before final machining is begun.

3.2 Many of the **components** of jigs and fixtures, such as nuts, bolts, levers and the like, are standard items in mild steel; but there are three groups of widely-used components which deserve special mention:

(*a*) Dowel pins, small shafts and spindles are usually made from silver steel, a material which can be hardened and is very hard-wearing.

(*b*) Jig bushes, which also need to be hard-wearing, are made from materials which differ according to the size of bush required. The smallest are made from silver steel, intermediate sizes from cast steel, and the larger ones from case-hardened mild steel.

(*c*) Non-metallic materials such as fibre, leather and plastics are used in clamping faces at all points where damage to the workpiece could be caused if metal-to-metal contact were allowed.

3.3 Jig bushes comprise that part of the drill jig through which the cutting tools—drills, reamers and boring bars—are located and operated. They fit into the **bush plate,** which may be made of either mild steel or cast iron. Precision of manufacture, selection of the correct type of steel, and strict observance of specifications for heat treatment—all are of such importance that most buyers nowadays prefer to buy only from specialist manufacturers.

Fig. 3.1 illustrates a number of standard jig bushes in general use, but it is often necessary to depart from standard sizes for an unusual job. **A** is a *stationary headless bush* such as would be used in cases where a projecting head would be inconvenient, as with the box-type jigs. **B** is a *stationary headed bush* which, like **A**, would be made to fit tightly into the jig plates. **C** is a *liner bush* which, as its name suggests, forms a lining in the soft material of the jig plate. It would only be used where the finished jig must have a long and accurate life. Liner bushes are also available without the shouldered portion.

D is a *renewable bush* used in conjunction with a liner bush, being retained in the jig plate by means of a small screw whose head fits the lip *b*. **E** is a *slip bush* which is held in position by the screw seen at **F**. Since slip bushes must be capable of being removed by hand, their heads are always knurled. They are useful when a small hole has to be drilled followed by a larger one, or for purposes of reaming. **G** illustrates a bush which needs to be of unusual length and shape to enable it to get as near as possible to the component which it is to guide.

When holes are grouped closely together, it may not be possible to

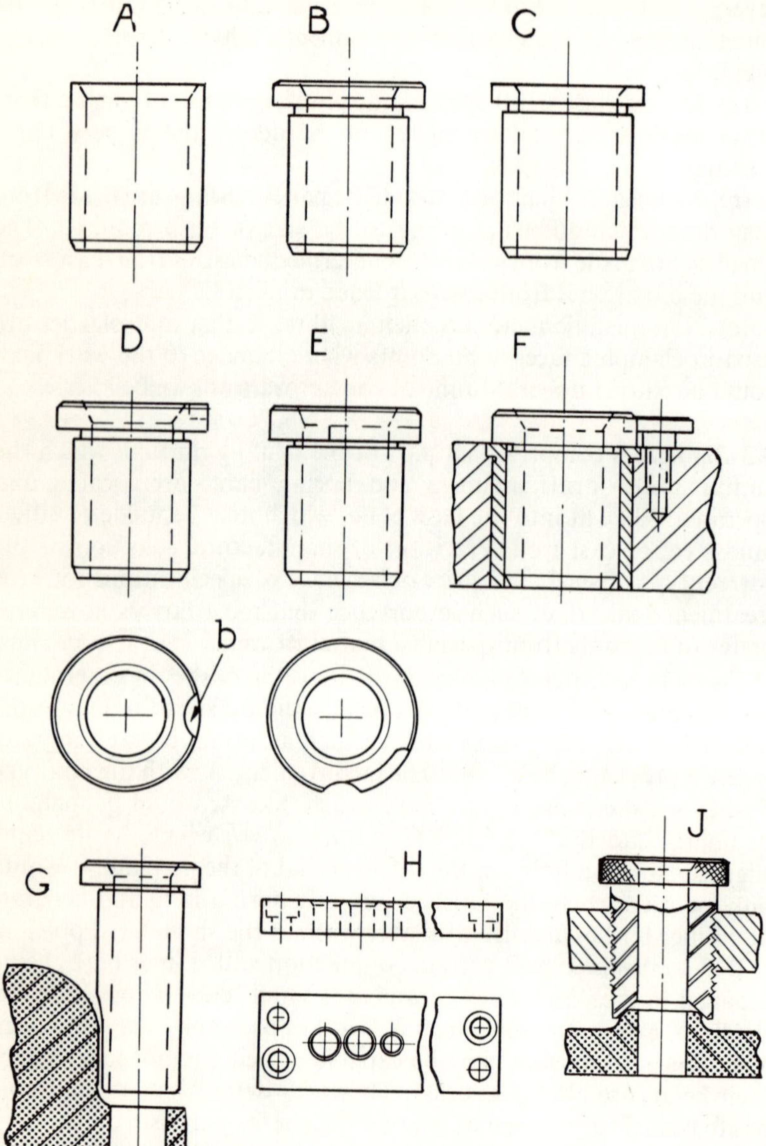

Fig. 3.1 A Selection of Jig Bushes

use any of the bushes mentioned, and steel plates such as that shown at **H** must be used instead. Such a plate must always be made to suit an individual job. It will be screwed and dowelled to the jig body.

Bush **J** is seen acting as a clamp and located over a rounded bossed component. Inaccuracy of machining will always result if there is any slackness in the fit of its thread.

With the exception of **G** and **J**, all the types of jig bushes mentioned are covered by B.S. 1098, which gives details of sizes, shapes and recommended materials.

3.4 Clamping Devices. The simple types of clamp illustrated in Fig. 1.3 are capable of being developed to make use of quick-acting devices, and to hold workpieces of different shapes. Fig. 3.2 shows a design which does away with the need for spannering and is useful where heavy clamping is not required.

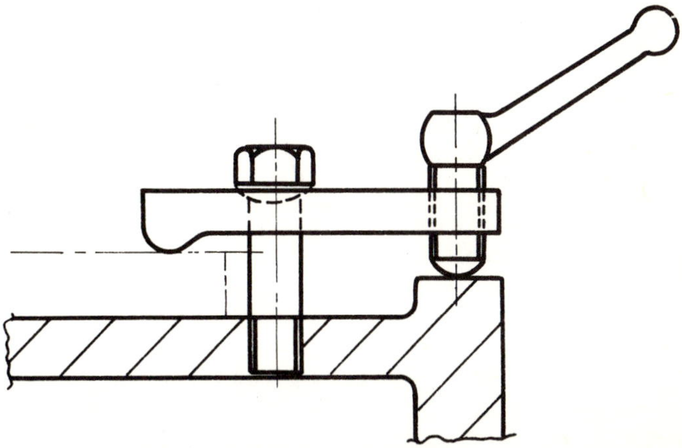

Fig. 3.2 A Hand-Operated Clamp

The spherical washer is fitted under the centre stud to enable the clamp to float to three-point contact between the work and the heel. The clamping lever is keyed to the screw, which in this case forms the heel pin. The point of the screw rides in a slot to prevent the clamp moving out of position.

Fig. 3.3 shows two arrangements of cam-activated clamps. This type of clamp can be employed to advantage when no great variation in the position of the component face to be clamped is likely. Clamping action is rapid, being achieved by a simple downward movement of the lever. The amount of eccentricity needs to be carefully determined. Cam rise must not be too rapid, and adequate mechanical advantage must be obtained throughout the stroke.

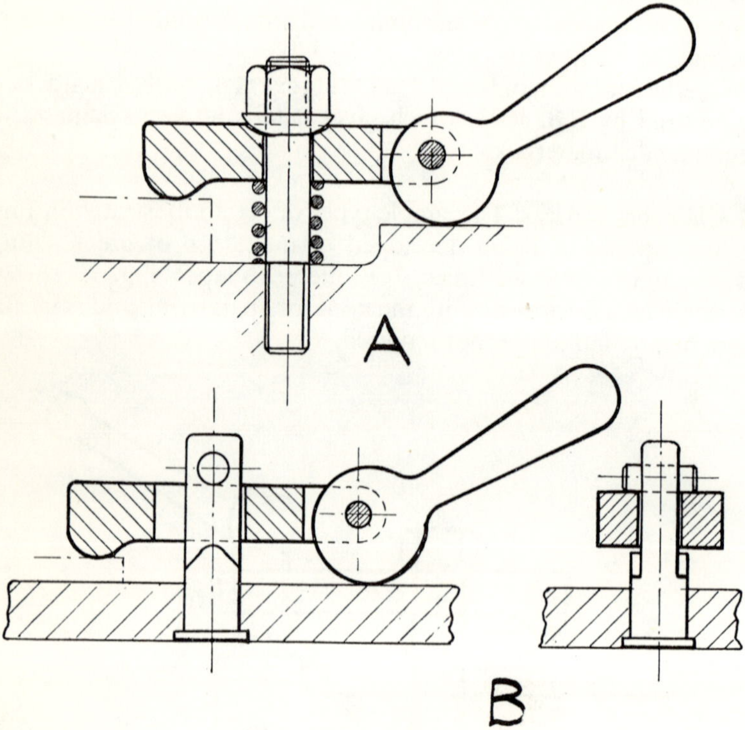

Fig. 3.3 Two Cam-Operated Clamps

Fig. 3.4 shows two examples of sideways, or *edge clamping*. This type of clamping is only employed when the face to be clamped is in practically the same relative position on each component after it is placed in the fixture. Nor will either of the clamps illustrated accommodate any great variation in length. Their advantage lies in the fact that the clamp itself does not project above the top level of the work, so that cutter travel is not hindered.

3.5 Indexing Devices. It is often necessary to be able to rotate or slide the component being machined from one fixed position to another—in order, for instance, to drill a series of holes with equal radial spacing. For this purpose use is made of *indexing devices*, of which a number of alternative designs are discussed below.

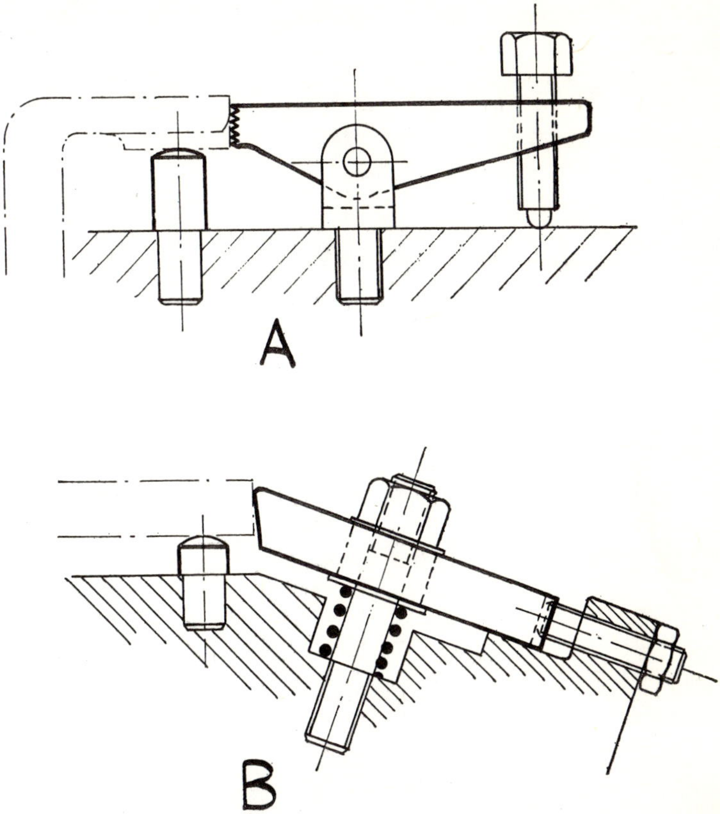

A

B

Fig. 3.4 Edge Clamps

The shape of the fixture and the type of machine being used will generally dictate the type of indexing device to adopt; but the latter should always be simple in design, easy to operate, and reliable. Fig. 3.5 shows, at **A**, a lever-operated design using material of rectangular section. Its spring can be located for either tension or compression, or a flat steel spring can be used instead.

The plunger-type indexing device shown at **B** is pulled out of its socket by means of the knurled knob until pin *b* is clear of the top plate. If the device is rotated in this position, the plunger will be held clear. In the type shown at **C**, the plunger is raised or lowered by cam action. When *b* is rotated, pin *c* rides up the inclined face of pad *d*.

Illustrated at **D** is the lever operation of a circular plunger; while **E** shows an alternative, and neater, design in which the spring itself is shielded.

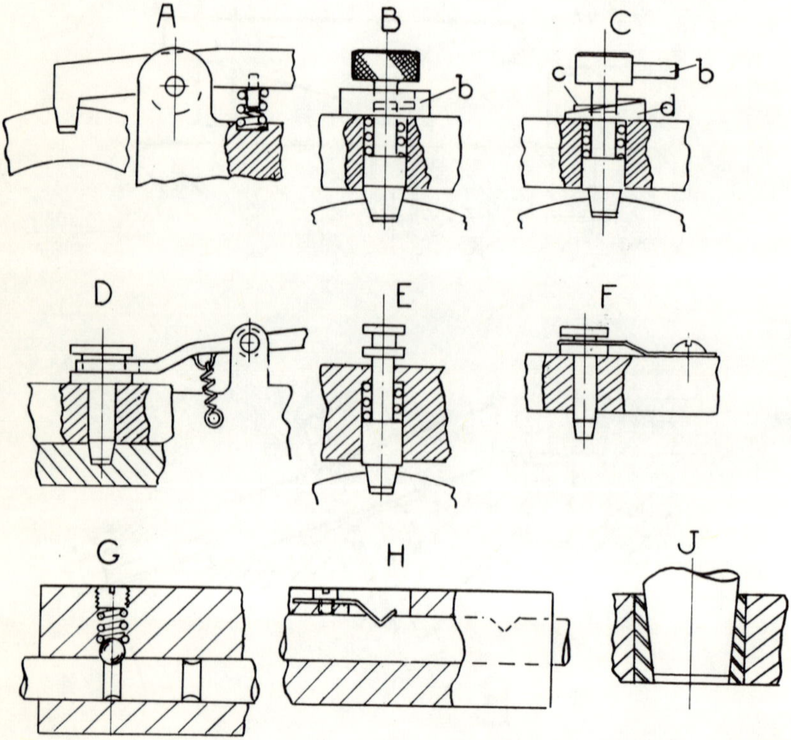

Fig. 3.5 A Selection of Indexing Devices

The designs illustrated at **F**, **G** and **H** are useful for the indexing of sliding mechanisms, particularly in cases where space is limited. In all of them both the plunger and its socket should be hardened to reduce wear, and there must always be a gap at the bottom of the socket.

It is a good plan to make the socket separately of tool steel so that it can be removed, as at **J**, and replaced when it becomes worn.

CHAPTER 4

Drill Jigs

It will be the purpose of this and succeeding chapters to consider the design of various pieces of machine-tool equipment, to introduce sample components, and then to design the equipment which would be required to carry out a typical engineering operation.

4.1 The different types of drilling jigs are best classified in accordance with the type of operation which each is designed to perform—the drilling of raw castings, of half-machined components, of parts which have been machined to coarse limits, to fine or to very fine limits respectively, and so on—but the reader is warned that the dividing line between types is often so faint as to be indistinguishable.

4.2 Locating Rough Components. Great care must be exercised in deciding from which point to locate a component, for on this decision the accuracy of the finished work will depend. At the same time excessive operations performed solely to provide a suitable locating point should be avoided wherever possible.

Once an operation has been performed on a piece, the question of location is simplified. It is with the rough component that the difficulty arises.

Forgings, stampings and die castings are generally produced uniform in shape; but ordinary castings vary considerably, and in dealing with them an elementary knowledge of pattern making or of foundry work is of great value. For it would obviously be fatal to locate from a point that was continually changing, and the knowledge referred to will enable this to be avoided.

Holes which are to be machined in castings are generally "cored" in the foundry, leaving an allowance of metal to be removed. But in the casting process these cores often move, so that the hole in the casting itself is unreliable as a locating position. The boss *surrounding* the hole, however, is produced by the pattern itself and can therefore be relied upon for locating purposes.

The correct location for a casting is best decided after examining several sample castings, which will reveal the method of casting used and should show up any doubtful points.

When locating a workpiece of irregular form, or when providing a seating for a component, a method known as *three-point location* is generally used. It is best explained by analogy, comparing a three-legged stool with an ordinary chair. Placed on irregular ground, the latter will always rock on the axis provided by two diagonally opposite legs, whereas the former will adjust itself in whatever conditions it is placed. The diagrams in Fig. 4.1 illustrate three different methods of locating components in their rough state preparatory to the first operation.

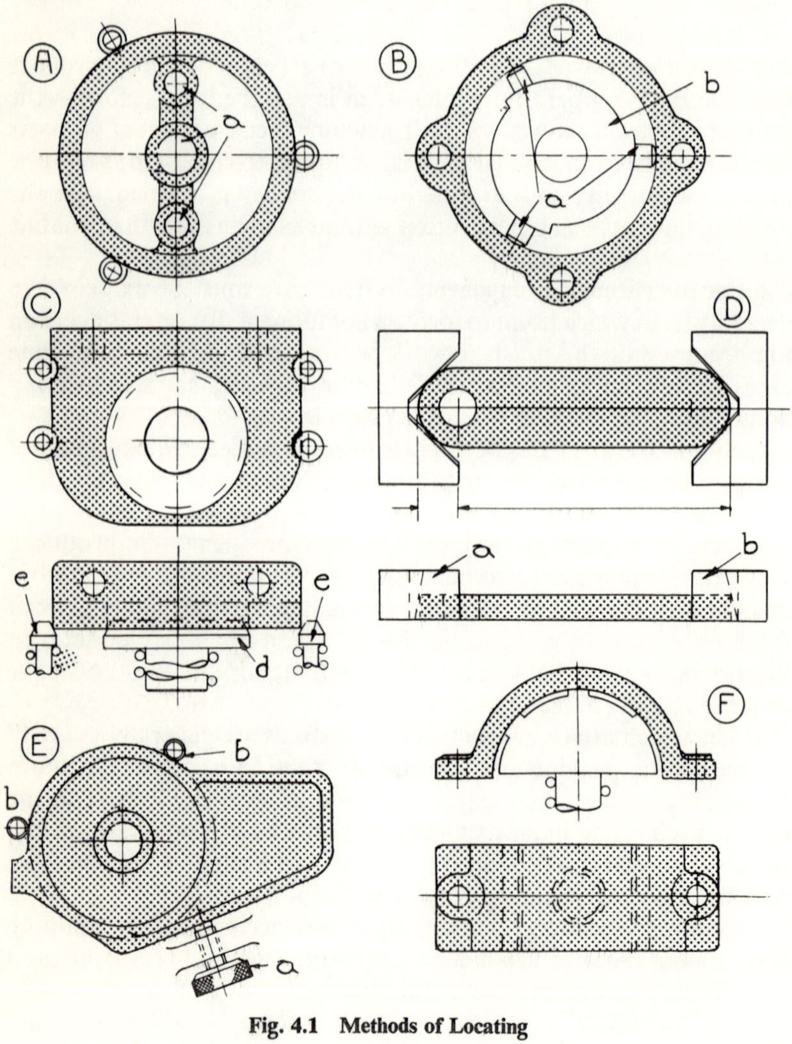

Fig. 4.1 Methods of Locating

The component at **A** is shown shaded, the operation required being to drill the holes *a* equidistant from the outer rim. In this case it is obvious that, even though the holes could have been drilled more central to the bosses by locating therefrom (which might have given the finished piece a better appearance), nevertheless the important point, *i.e.* the relation of the holes to the outside rim, would have been missed, and a quantity of work would probably have been wasted in consequence.

One form of location is shown at **A**, where three coned studs are provided equally spaced round the rim at such a radius that the rim seats on the cones. The latter have springs mounted beneath them, which enables the cones to drop or lift with the movements of the component. The latter is by this means centralised, and the drill bushes need only to be fixed on the jig in correct relation to the coned pegs to ensure accurate results.

The angle of the cone on the pegs is generally made about 60°. They can be arranged to bear on either the inside or the outside of the component according to choice.

Another form of location is shown at **B**, where the rim is again important. Here, correct location is secured by three hardened steel plungers *a*, behind which are fitted springs. When these plungers are actuated by a partial rotation of cam *b*, which forces them out equally, the piece is accurately located. The springs cause the plungers to fall back automatically once the cam is released.

The cast component illustrated at **C** has a cored hole. It is required that two holes shall be drilled through its flanges approximately central with the bore and parallel with the sides. Had the bore been machined, it would have been a simple arrangement to slide the work on to a short plug and to clamp it with a nut and "U" washer. This being impossible, the difficulty can be overcome by mounting a coned centre *d*, capable of a sliding motion and fitted with a spring, and four coned pegs similarly fitted at the sides. When the component is pressed down on to the coned centre, the latter gives way and the spring is compressed until the component comes in contact with the coned studs *e*. These studs set the sides to the vertical, after which the piece is clamped.

There exist a large variety of levers and plates which require to have holes drilled through them a certain distance from the end. A form of location for this operation is shown at **D**. When position of the hole in relation to one end of the plate only is important, the correct dimension will be obtained if the work is located with the end which is to receive the hole placed in the fixed vee of the jig, the other vee being used for clamping. Since the work may tend to lift when it is clamped, the pads *a* and *b* are often provided with a chamfer as shown, to help in holding the work down.

For short components, special jaws fitted to a machine vice could be used quite satisfactorily for this operation.

Parts having an exceptionally irregular outline are difficult to locate. One method of doing so is shown at **E** in Fig. 4.1. The component is forced by the screw *a* against the two pegs *b*, the latter being arranged to bear against the most clean and reliable parts of the casting—points which also govern the important dimensions.

When the piece is so irregular in shape that it is impossible to use pegs in this way, the difficulty has been overcome by accurately marking or engraving the outline of the piece on a jig plate, and then clamping the piece on to the plate over the outline. A skilled operator will be needed for a job of this kind, but the extra expense will generally be outweighed by the small amount of scrap produced, and by the time saved by not having to adjust a jig with pegs every time a different casting is to be located for accurate machining.

Illustrated at **F** in Fig. 4.1 is an example of a case where the exact positioning of a radius is important. For this purpose a semi-circular pad provided with a spring seating such as that shown works quite successfully. To overcome the roughness of the concave shape, the pad is cut away or relieved so that only two or three points of it are left in contact with the component.

4.3 Clamping Devices. With the component correctly located, it must next be efficiently clamped. In Fig. 4.2, **A** illustrates a type of clamping which is much used, particularly in drill jigs where drilling has to take place from a previously bored hole. The component is pushed over a locating pin, and the clamping plate is swung into position under the nut. In this way the plate cannot be lost, for it remains fast to the fixture—a feature which makes this method of clamping much to be preferred to a slotted washer, which can easily disappear.

The device shown at **B** is useful for rapid clamping, especially when machining time is short and handling time must be kept to a minimum. It is only suitable for light clamping, however, and then only in cases where the components are of approximately the same length. The clamping lever is hinged on the clamping bar, and this in turn is hinged on to the jig body. Though rather more expensive to manufacture and maintain, the cost of this type of clamping is quickly recovered from savings in handling time when the number of pieces in a production run is large.

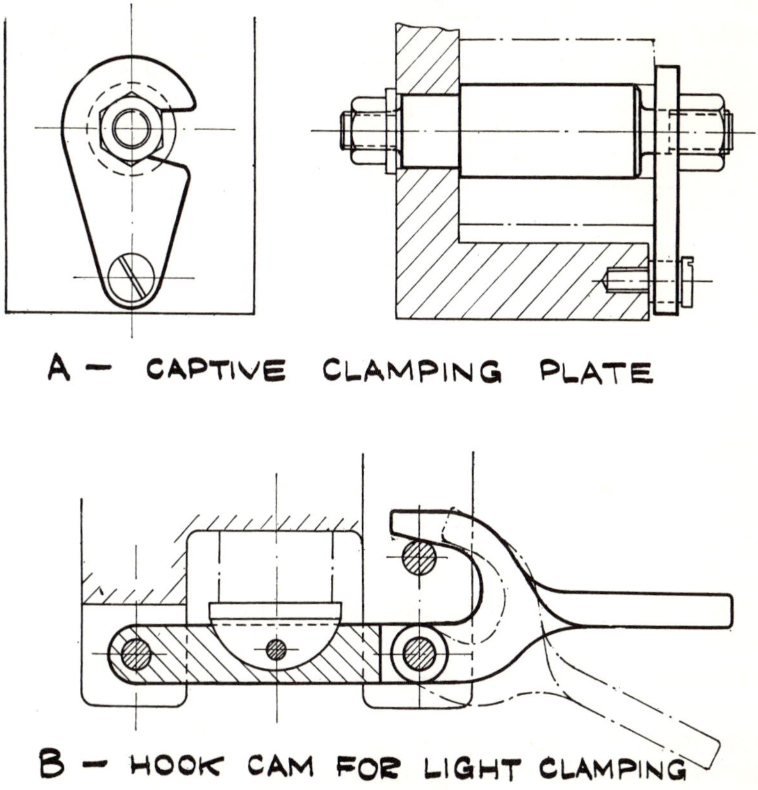

Fig. 4.2 Two Clamping Devices

Fig. 4.3 shows a design of differential clamp, which has advantages when the component is loaded from one side. This type of clamp will adjust itself to errors in component shape, and is sufficiently robust to absorb a heavy cutting load while at the same time giving adequate support. The clamping action is as follows: Clockwise rotation of the star wheel *a* thrusts the wedge bar *b* forward so that it advances to side clamps *c* until the end of the travel of the slot is reached. Further rotation of the star wheel then causes the jaws to close, and the side first to meet the component is held up until the second jaw makes contact. At this point both clamps are fully differentiated.

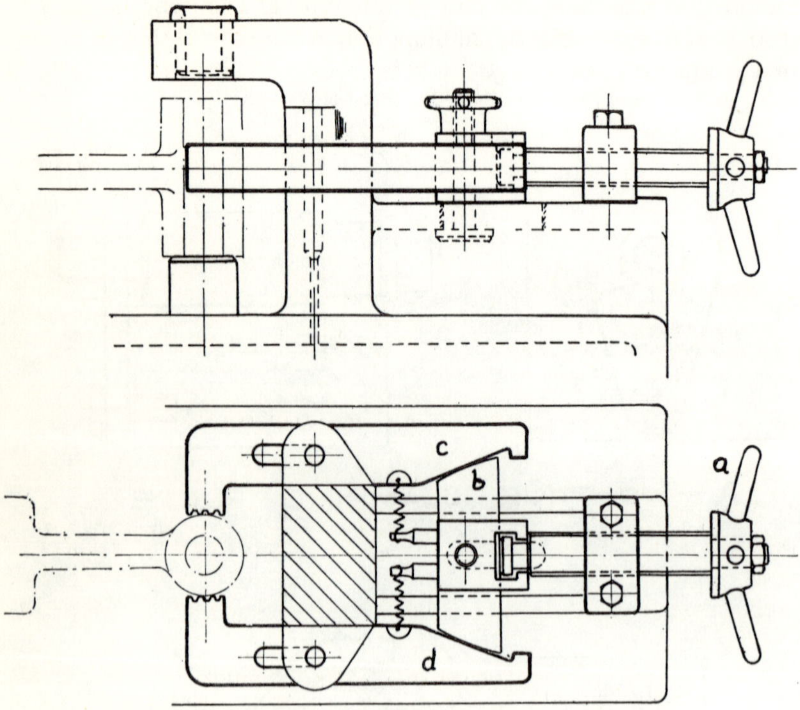

Fig. 4.3 Differential Clamping

4.4 Simplicity of Design. It has already been said that mass production does not necessarily call for elaborate and multiple tools. Sometimes, the simplest equipment will give the best results. It is possible, for instance, to design a single drill jig which includes bushes for every hole in the component. But while this method is safe in that it ensures an exactly correct relation between holes, production can be hampered if it is necessary to wait for one operator to complete the drilling. Had a more simple jig been designed, the work could have been split among several operators, and overall production time could perhaps have been saved.

Fig. 4.4 shows a group of drill jigs such as have been used for drilling the necessary holes in a motor-car windscreen bracket. The components in question required a number of holes to be drilled in various planes. To have put the drills for all these holes in one jig would have resulted in a very awkward piece of work, so it was decided to make a number of simpler jigs as shown.

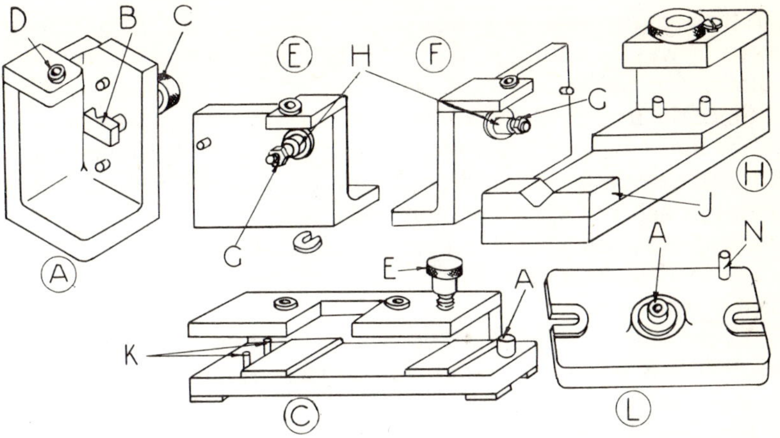

Fig. 4.4 A Group of Simple Jigs used for Drilling a Motor-Car Windscreen Bracket

The two brackets pictured in Fig. 4.5 were produced in pairs, left-handed and right-handed. For the jigs, angle-shaped iron castings and bright mild steel plate (the latter unmachined) were dowelled and screwed together.

The purpose of jig **A** was to drill a hole through the lug *f* (Figs. 4.4 and 4.5). For this purpose the component was held vertical by means of the hook-shaped clamp B operated by the nut C and located from the holes *e*. (It might have been found an advantage to fit a key in the shank portion of **B** which, by engaging in a keyway

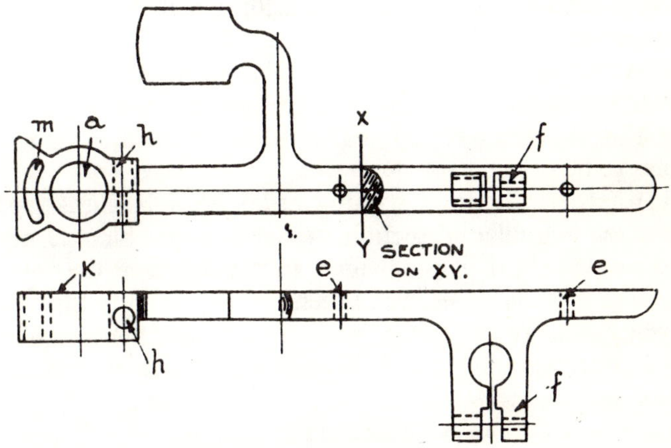

Fig. 4.5 The Windscreen Brackets to be Drilled in the Jigs shown in Fig. 4.4

cut through the jig casting, would have been prevented from turning when C was operated. A Woodruff type of key would have been suitable for such a job.)

Since the hole in the lower fork of lug *f* was tapping size and that in the upper fork was clearing, bush D was made to accommodate the latter so that, when drilling began, the clearing drill was set to pass through the upper lug and make a centre hole in the lower. Drilling was completed with a tapping-size drill, starting in the centre made by the larger drill. This necessitated the use of a two-spindle drilling machine.

As the larger of the two holes was only 6 mm diameter and the material was very soft, no support was necessary underneath the lug to take the thrust of the drill.

Jigs **E** and **F** were used for drilling the hole *h*, one being left-handed and the other right-handed to suit the opposite-handed components. They are of simple construction, being plain angle-iron castings. The components were located from the hole *a* on the stud H, the machined face *k* of the component butting against the shoulder seen on the stud. The latter was of hardened steel, was a push fit (see Chapter 11), and had a nut for locking purposes. To facilitate loading and unloading, nut G was made of smaller diameter than hole *a*, so enabling the component to be removed without taking the nut from the stud. The U or clip washer seen at the foot of the jig allowed the clamping to be effected.

These jigs were bolted to the machine table, so a quick-change drill chuck was used to enable the same drilling methods to be used as were employed for drilling the holes in *f*. By substituting tee-section angle for the jigs bolted to the table, the substitute jigs could have been made movable. Note that, as a general rule, the feet or foundation of a jig should be equally balanced about the area bounded by the holes to be drilled.

Jig **H** was for drilling and reaming the large hole *a*. The component, which is semi-circular in section, rested in a vee block J and was located by two short pegs engaging in the slot *m* at the top of the component. No clamping was necessary because the thrust of the drill, which was about 25 mm in diameter, kept the component in position. For reaming, the drill bush was removed (having been specially designed to facilitate this operation) and a bush of reamer size was inserted. The bush and its locking screw are seen in the illustration.

Note that bushes of this type, in addition to the clamps and nuts and various other parts of jigs, are standardised in different drawing offices. Needless to say, the designer should take care to follow any procedure laid down.

Jig **C** was used for drilling the two holes *e*, plug A accepting the large hole *a* of the component, which was also located by the two pins K. Clamping was effected by means of the knurled nut E, though a careful and experienced operator would not have needed such a precaution.

The jig itself was equipped with feet to keep it clear of swarf. When the area of a jig is large, such feet are essential even if the jig is clamped to the machine table.

The face *k* of the component was finished by an operation known as *spot-facing*, carried out on the plate L. Hole *a* fits plug A on the plate, with the pilot pin of the cutter a running fit in the small hole seen in this plug. The purpose of peg N was merely to prevent the component from swinging round horizontally, which could be dangerous when the cutter is at work. The plate itself was of cast iron, and was provided with two slots cast in it for fixing purposes. The centre plug was of hardened steel, a force fit (Chapter 11) into the plate.

4.5 Pot Jig. A jig of this type is illustrated in Fig. 4.6. It could be conveniently used for the drilling of small circular components which can be located at both inside and outside diameters.

The jig consists of a bush plate **A** (which also provides the locating spigot) and of the pot B, which gives the necessary support during drilling. Note that the pot has slots coincident with the drill bushes to allow clearance of swarf and burrs pushed through by the drill. For the same purpose a series of holes is provided at the bottom of the recess.

The locating pin **C** fits loosely into a slot in the edge of the bush plate.

Clamping is achieved by means of the captive "C" washer **D**, which is held loosely by the shouldered screw **E** to prevent loss. The washer is swung aside after freeing the clamping nut so as to allow the nut to pass through the central hole in the bush plate.

4.6 Box Jig. Some of the most awkward components which require drilling are the great variety of small brackets used in most branches of engineering. Often the holes to be drilled lie in two or more planes, and for this application a box type jig is used.

The example shown in Fig. 4.7 is of a second-operation drill jig, the three holes in the base having been drilled previously. Clamping is effected by means of the strap **A**, which carries a compensating pad **B** to ensure even pressure. The strap is locked by means of a hook

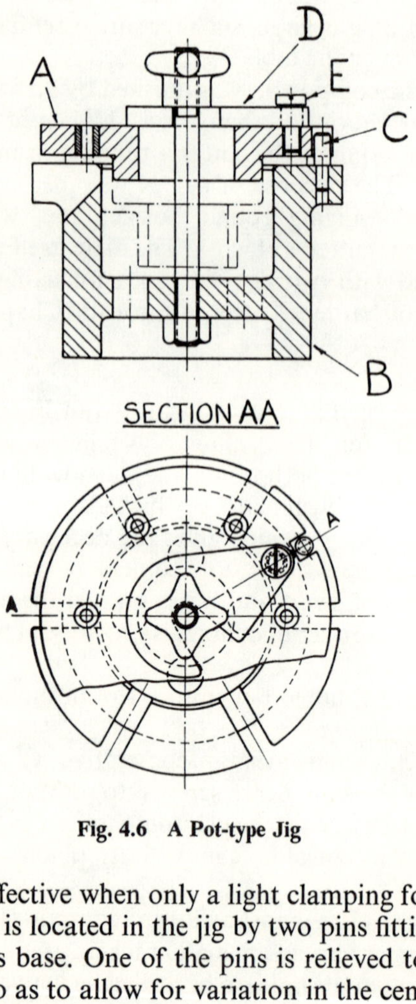

Fig. 4.6 A Pot-type Jig

cam, which is effective when only a light clamping force is required. The component is located in the jig by two pins fitting into holes in the component's base. One of the pins is relieved to locate on one diameter only so as to allow for variation in the centre positions of the drilled holes while still providing positive location.

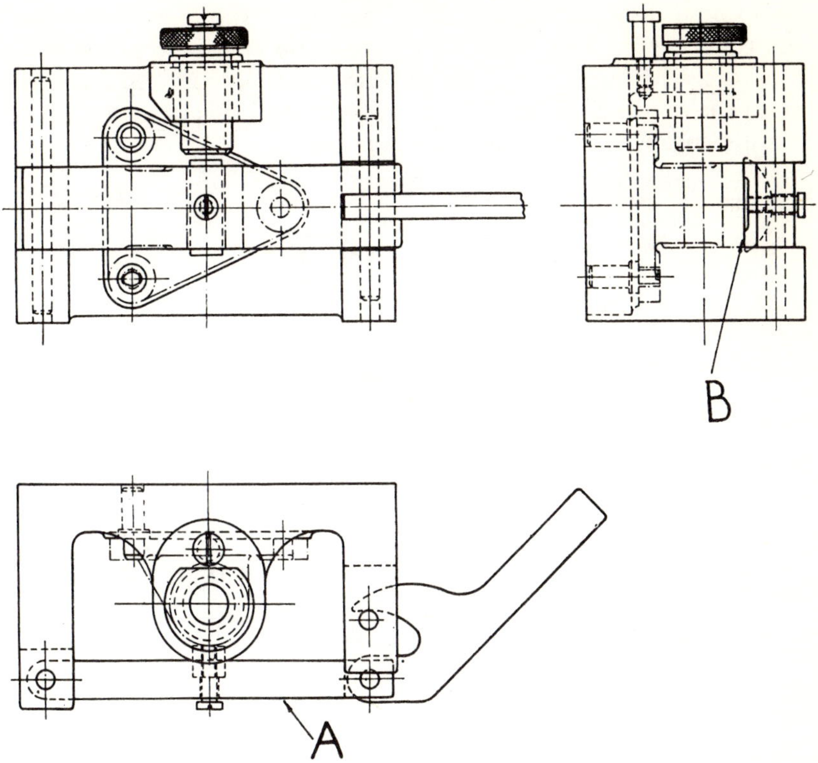

Fig. 4.7 A Box-type Jig

4.7 Jigs for Pistons. Fig. 4.8 shows a piston in section, with another in position inside the drill jig also illustrated. To get pieces such as these machined quickly, a horizontal machine having two spindles opposed to one another and capable of being fed in simultaneously is often used. Drilling time is halved, and by drilling through from opposite sides greater accuracy is achieved. For the piston to function properly when finished, it must be correct in both size and weight, and must have walls of sufficient thickness to give the necessary strength.

To secure the required evenness of wall thickness, special precautions are taken when the diameters are being turned (see Chapter 6). For greater ease when putting in the grooves, and also to provide a locating place when drilling, the inside diameter *a* is machined to rather close limits. The hole to be drilled, which is to take the gudgeon pin, must be central, at right angles to the axis of the casting, and at an accurate distance from the top of the casting. If this latter distance varied, it would alter the compression ratio when the piston was working. The face *c* and the diameter *a* are turned when locating from the top *d*; so if the hole about to be drilled is located from *c* the dimension *b* must be maintained correctly.

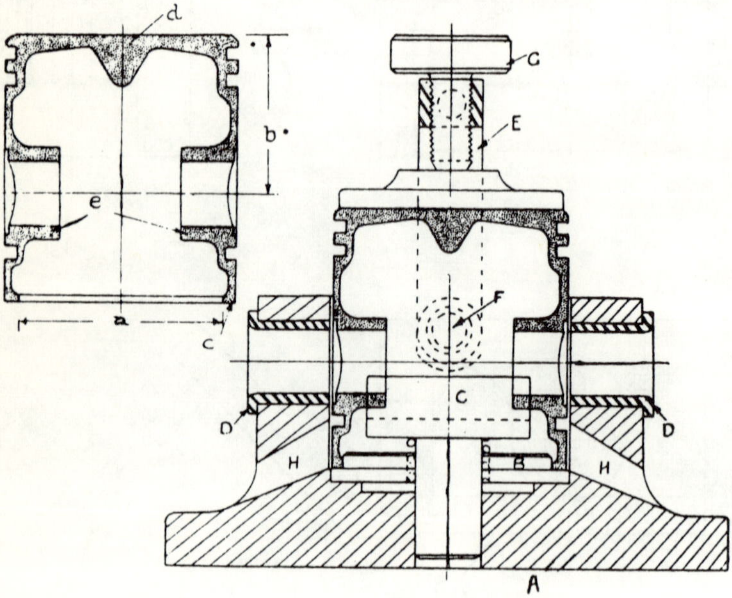

Fig. 4.8 A Horizontal Duplex Drill Jig for Machining a Piston

The jig illustrated thus consists of a main casting **A** which is clamped to the bed of a horizontal duplex drilling machine. **B** is a hardened steel locating pad, and **C** is a 60° vee block mounted on a peg and capable of a vertical sliding motion. It is provided with a key to keep it in position, however, so that the vee forms a seat for the two bosses *e* and locates them central with the drill bushes **D**. Beneath the pad is a helical spring which presses the pad upwards, thus assisting location. The clamp **E**, which swings about **F**, locks the casting securely in position. With so delicate a casting, the locking nut **G** must not be allowed to provide too much leverage lest the casting be distorted.

Any form of jig resembling a box or other receptacle naturally poses the difficulty of cleaning or otherwise getting rid of the cuttings which, if allowed to remain, would impair the accuracy of the jig. This is a point deserving careful planning, for successful cleaning should not occupy anything approaching the time taken for the actual drilling. At **H**, therefore, two outlets are cored in the body casting just beneath the bushes, such that the falling swarf will work its way out of the jig and any cuttings that may accumulate in the vee piece are easily swept out in the same direction.

To prevent, as far as possible, any cuttings falling on the main location **B** and under the piston, the former is relieved so as to leave only three points for the latter to rest on. The space between the vee block and the pad should also be kept deep enough to prevent swarf interfering with its free movement.

4.8 Jigs for Use on Two-Spindle Machine. Fig. 4.9 shows a jig used for drilling the darker-coloured component seen in position inside it. This component (it happens to be part of an agricultural machine) is a steel forging which has undergone its first operation. This operation consisted of boring the centre hole, and then facing boss *b* of the latter and also bosses *a* and *c*. The two holes *x* and *y* have still to be drilled.

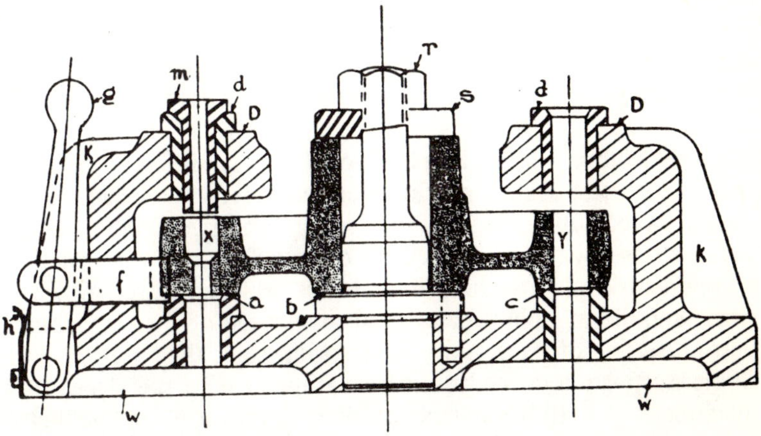

Fig. 4.9 A Jig for Machining a Component in a Two-Spindle Machine

The most important point to be ensured in the drilling is that the holes are drilled the correct distance apart, and that they run parallel with the centre bore. Since the latter is finished and faced, as are also the two small bosses, the method of locating shown would be adequate to give the desired result.

The jig illustrated is intended to have a long life, so the seats on which the bosses rest are made of hardened steel. The arms *D* carrying the drill bushes *d* should be positioned as close to the forging as possible so as to minimise the length of the drill protruding below the bush. This reduces the risk of the drill deflecting from its true course. To locate the bosses *a* and *c* central with the bushes *d*, the vee block *f* is operated by the lever *g*. The spring *h* is used to keep the block in contact with the component when locking nut *r* is secured. The height of *f* should be such that it touches the forging just below the centre line of the boss. In this way the "flash" or burr which often disfigures forgings because they have been formed between two half dies will be avoided.

Note that the cast-iron arms bearing the drill bushes are strengthened by the webs *k*, which add no excessive extra weight. (Though webs, fillets and other strengthening devices must often be included in the design of a fixture, they inevitably increase its cost by reason of the extra work involved in the making of the foundry patterns.)

With the jig shown in Fig. 4.9 mounted in the double spindle machine, operational procedure is as follows: Boss *c* is drilled right through, but boss *a* only half-way through—after which the small bush *m* is slipped in position and the hole is drilled right through. Had both holes been of the same diameter, the job could have been done on either a single-spindle or a multiple-spindle machine. The latter would of course have taken only half the time for the job.

4.9 The Multiple Drill Head. Though many different designs of multiple-spindle drilling machines are in current use, recent developments have emphasised the design of multiple-spindle heads of various sizes, with spindle centres designed to deal with particular components and with the spindle heads attached to radial or other single-spindle machines. The great advantage of these machines is that no setting of the spindle is necessary, since they are made for a single purpose and need only be clamped on to the sleeve of the machine. Fig. 4.10 presents a sectional elevation of one having two spindles which would be suitable for use with a drill jig of the type shown drilling the two holes at once.

The body is built in two parts, **A** and **B**, of cast iron which should not be too heavy. The use of thin section with ribs introduced to add strength is recommended for the purpose. The drill spindles **C**, which are of mild steel, are integral with the pinion by which they are driven. They are run in and supported by the phosphor bronze bushes **D**. The upward thrust of the spindles is taken by the ball thrust washers **E**, which are dropped into a seating in the spindle pinions and are

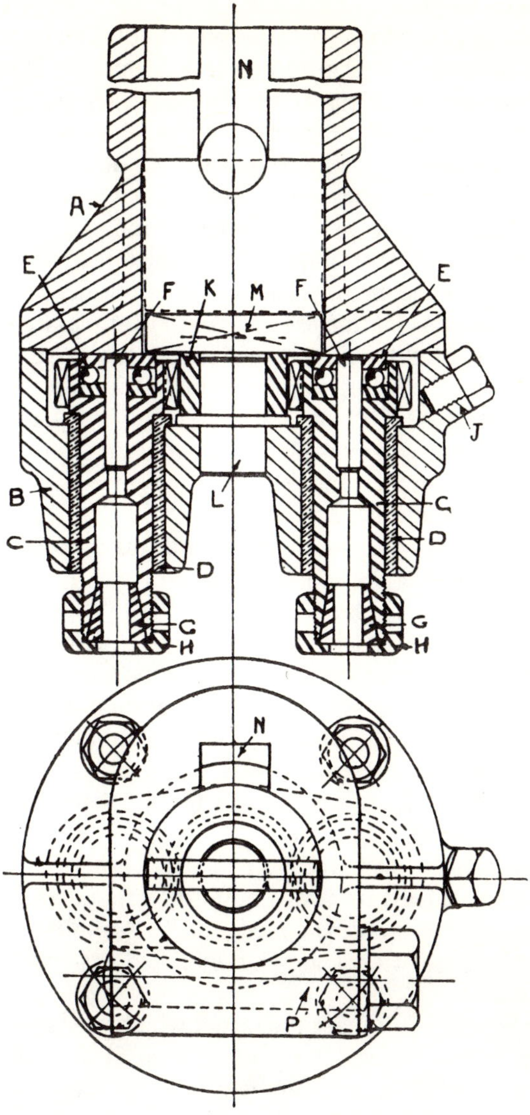

Fig. 4.10 A Multiple-Spindle Drill Head

further retained by the rod **F**. The drills are gripped by the spring collets **G**, which are forced up into the coned portion of the spindles by the nut **H**. The latter has either holes or flats to facilitate tightening.

The gears are lubricated through the plug hole **J** (though if they are packed with good solid lubricant during assembly no further attention should be necessary for a long time). The two spindles are driven by means of the central pinion **K**, which runs on the peg **L** and is in turn driven direct by the drilling-machine spindle. A slot some 10 mm square is cut centrally across the end of the machine spindle and engages the square peg shown at **M**, the latter being fixed centrally across the centre pinion. Alternatively, a tapered spindle is made integral with the centre pinion and secures the drive by fitting into the hollow spindle of the machine.

The complete attachment is slipped on to the sleeve of the drilling machine, with the rack attached to the latter fitting into slot **N**. When the bolt **P** is then tightened, the head is securely held.

It will be seen from this simple example that it is not difficult to design a different head for any arrangement or for any number of holes. If the holes are to be drilled closer together, the gears will have to be of smaller diameter. In Fig. 4.10, the centre pinion had 19 teeth and the two side pinions 17 teeth each.

The spindles can, if desired, be bored to take morse taper shanks instead of the small collets and nuts shown.

4.10 Indexing Jigs. Fig. 4.11 shows the type of drill jig which would be used for a component such as the centre portion of an automobile rear axle. The holes to be drilled are six in number, equally spaced round the flanges *a* and *b*. A horizontal multi-spindle drilling machine not being available, the fixture was designed to be used under a vertical machine such as that shown in Fig. 4.14, from which the tables were removed before use.

The component was located from its cored hole, the flanges having been previously turned from this point on a mandrel similar to that shown in Fig. 6.2 and on the coned stud **C**. A similar stud $\mathbf{C}_1$ at the opposite end, fitted with a screw, completed the location and also acted as a clamp.

In order to set truly square the face *f*, which was about 250 mm in diameter, a rod with cams was arranged at **H** in such a way that, before $\mathbf{C}_1$ was tightened, the cam was brought into contact with the face of the piece and held it in position while the clamping was effected.

The fixture itself consisted of a cast-iron cradle, kept as light as possible and strengthened by webs and fillets at **M**, revolving on steel spindle **N** in the pedestal **P**. The latter had a bronze bearing, which was made as long as possible. The webs **R** and **S** supported the

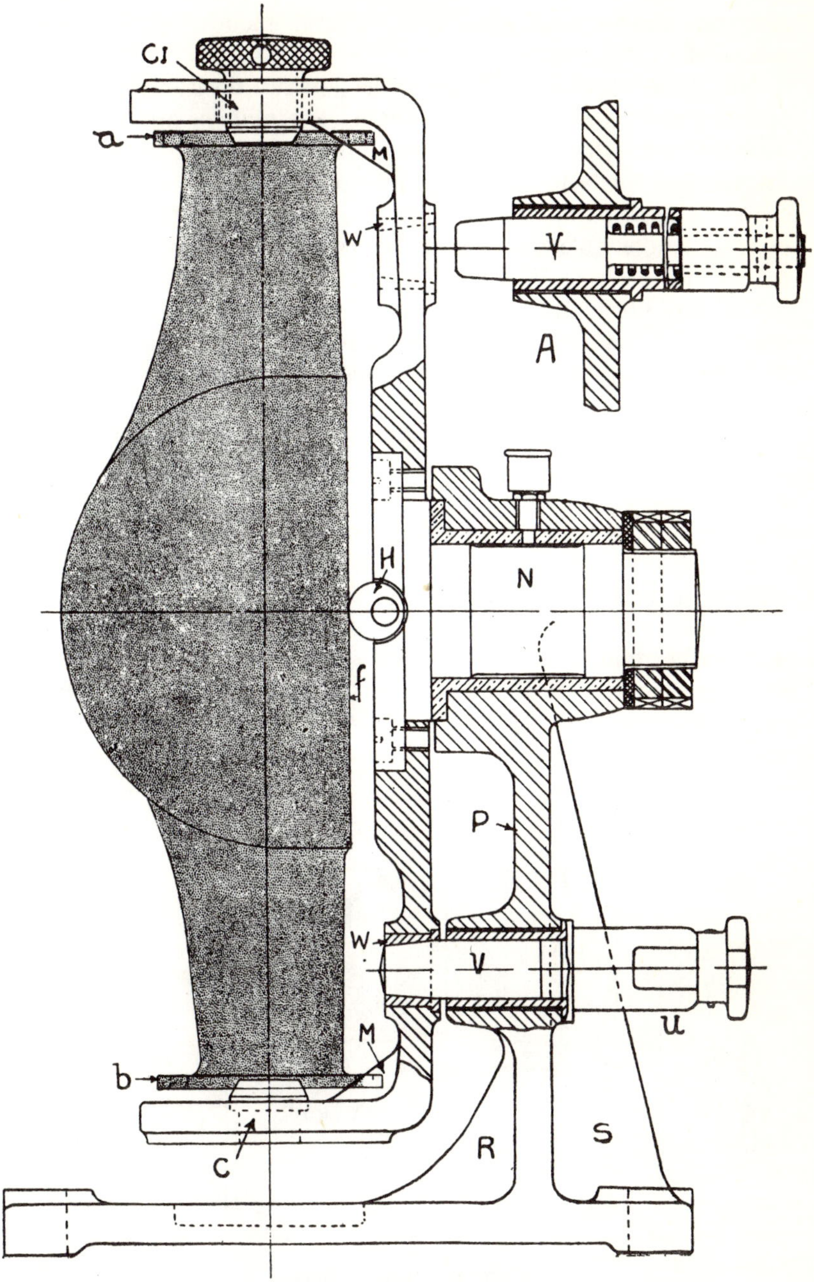

Fig. 4.11 An Indexing Drill Jig

pedestal, which was subjected to considerable load since all six holes were drilled at the same time.

The plunger for indexing the cradle in its two positions is shown in position at **U**, and also in detail at **A**. The plunger pin **V**, and the bushes **W** into which it fitted, were of hardened steel to prevent wear. The plunger pin itself was made rather long so as to prevent it working slack—with resultant risk of faulty indexing. It was given a tapered nose so as to enable any wear to be taken up by allowing it to drop deeper into the bush **W** whenever indexing action was required.

Fig. 4.12 illustrates another type of indexing jig, which could be used on either a radial or a multi-spindle drilling machine (see Fig. 4.14)—the component to be manufactured being in this case an automobile gear-box casting. The jig is seen in position on an ordinary radial drilling machine. Since only one hole can be drilled at a time unless a head designed on the principles explained in Section **4.9** be employed, production on such a machine will be limited. This drawback could be overcome by mounting the jig on a steel truck and passing it beneath a series of drilling machines, each set to operate on a different face of the casting.

The main cradle of the jig is of cast iron, and pivots on steel studs set into the standards **A**. It has four steel indexing bushes, three of which are visible at **B**, **C** and **D** respectively. The plunger for them is seen at **E**.

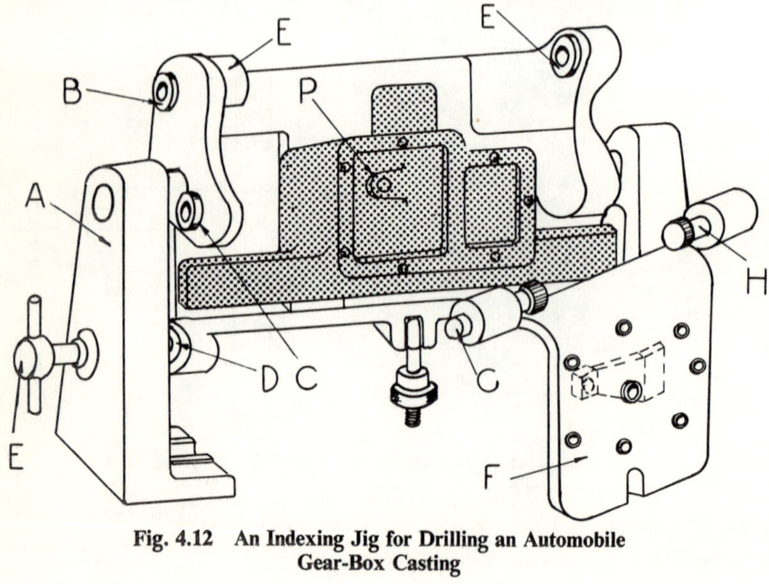

Fig. 4.12 An Indexing Jig for Drilling an Automobile Gear-Box Casting

To enable the component to be easily loaded and removed, the top plate **F** is made so that it can be detached from the jig. To hinge this plate as if it formed the lid of a box would have necessitated widening the cradle, which would have been undesirable because the drill bushings must be kept as near as possible to the component. The plate was therefore located between the bushes **E**, into which the plugs **G** and **H** were then pushed.

It is always desirable that jig bushes should be placed as close to the workpiece as possible, so the presence of a lug inside the gear box well away from the bush plate presented a difficulty. To prevent the drill from "wandering", support near the point of entry must always be provided. The way this was done is illustrated in Fig. 4.13, where **A** is the lug to be drilled and **B** the bush through which the drill will enter: the bush **C** is attached to another removable part of the jig and acts as a support. The plate **F** pictured in Fig. 4.12 has on its other side an arm carrying a supporting bush which is clearly seen in Fig. 4.13.

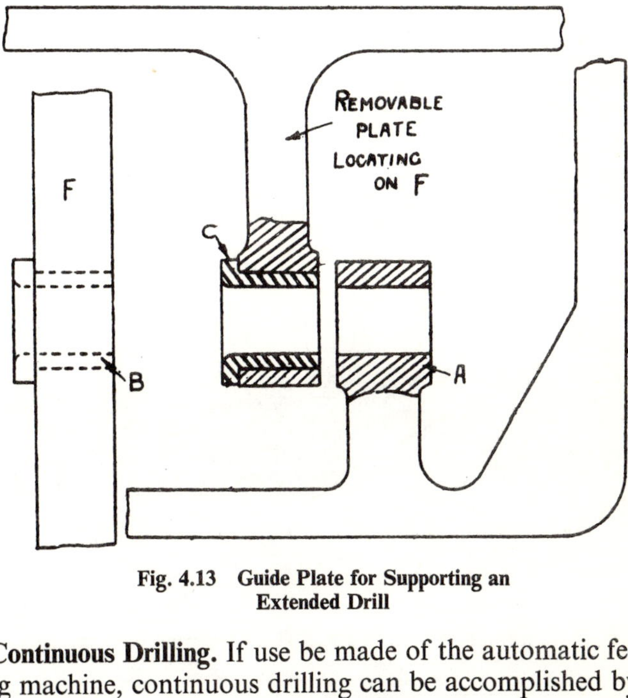

Fig. 4.13 Guide Plate for Supporting an Extended Drill

4.11 Continuous Drilling. If use be made of the automatic feed of a drilling machine, continuous drilling can be accomplished by using two identical jigs located on a revolving plate arranged on the machine table. Fig. 4.14 illustrates an example of this technique in which components for a harvesting machine are being drilled.

The two fixtures needed are seen built as one, so that while one is in operation the other can be loaded. The jig plate **A** carrying the drill bushes is suspended on four steady rods in such a way that the plate follows as the spindles are brought down to or are lifted from the work. Correct locating of the plate on the jig is assured by means

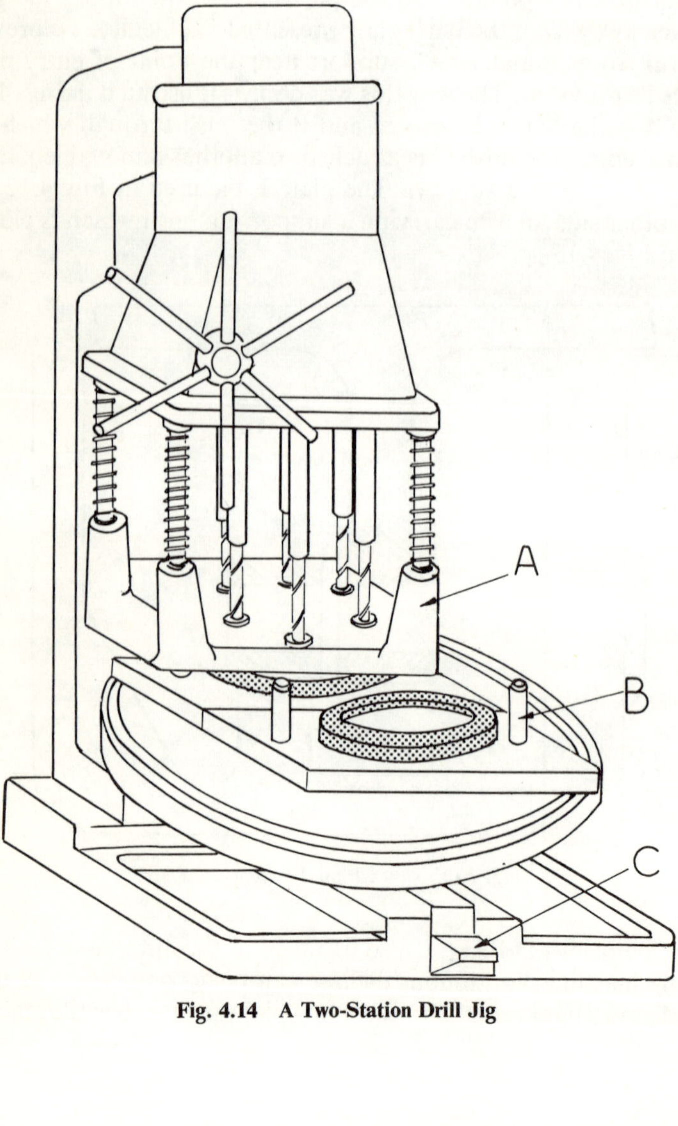

Fig. 4.14 A Two-Station Drill Jig

of the pegs **B**. The purpose of the spiral springs seen on the steady rods is to keep the plate in position at the bottom of the rods when the spindles are lifted after drilling.

For a reaming operation, three jigs equally spaced round the table could be employed. The machine spindles would be arranged in two groups, one holding the drills and the other the reamers, in such a way that while one component was being drilled, another was being reamed and a third was being loaded into the jig.

The revolving table itself is generally purchased with the machine, but it can easily be made separately. The indexing plunger is seen at **C**. To save time and more effectively to remove the cuttings, a compressed air pipe can be attached to blow away the swarf as it accumulates.

CHAPTER 5

Milling Fixtures

5.1 The finishing of surfaces can generally be performed in any one of three alternative ways—on the milling machine, on the planer or on the grinder. When the work is other than a plain surface, the milling machine is normally chosen (though the decision to use it is seldom automatic).

The next question to arise is whether the fixture to be used is to be a multiple one holding a number of pieces at once, or one designed to operate on a single piece at a time. The answer will depend on the number of components passing through the shop, and on whether the operation is one which is likely to be repeated on receipt of further orders at a later date.

5.2 Choice of Machine. The selection of the type of milling machine most suitable for a certain operation is quite often a matter of machine availability, a job being put on to a certain machine because another and more suitable one happens to be currently full of work. Nevertheless, the designer should always consult the shop foreman as to the best machine for any job, paying particular attention (especially in the case of "string" fixtures) to the capacity or length of traverse of the machine table of any machine that may be available.

A milling machine of the "Plano" or "Ingersol" type is undoubtedly valuable for medium and heavy work, especially when large quantities of components are available; but this type is not known in every shop, so more common machines should also be considered. The points to be watched by the designer do not differ greatly from one type of machine to another. When a component requires plain surfacing, the choice usually rests between horizontal and vertical milling (see **A** in Fig. 5.1). If the surface is very wide, it will probably be quicker to use slab mills on a horizontal machine rather than an inserted tooth cutter in a vertical machine. On a small area, however, a vertical operation will nearly always be the quicker.

The component illustrated at **B** calls for "straddle" milling. For it, the fixture could be made adaptable so that if the horizontal machines are full of work, the fixture could be arranged on an angle plate as at **C** and vertically "straddle" milled. (Since the end of the arbor is unsupported, though, a slower feed would in some machines be

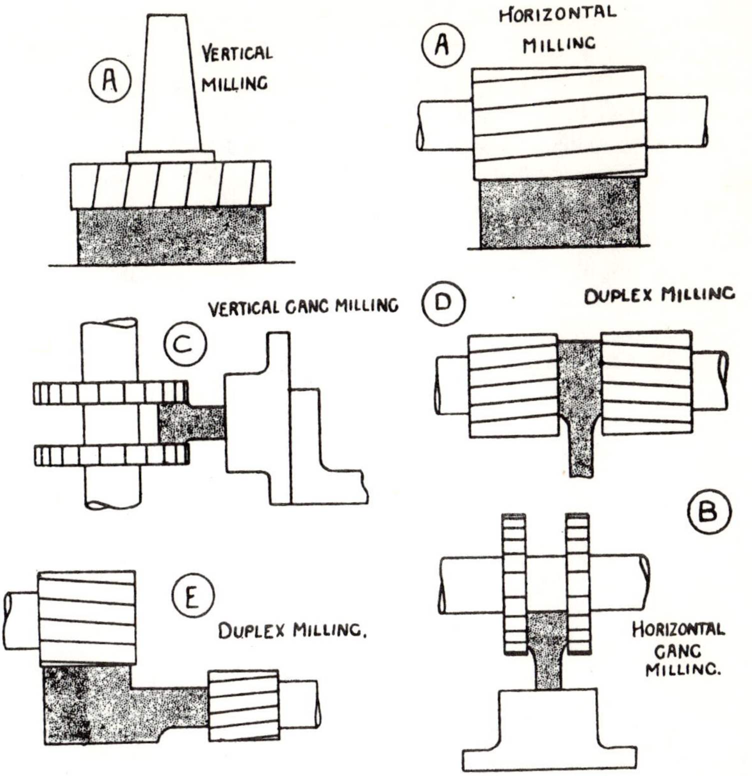

Fig. 5.1 Some Milling Methods

necessary.) Alternatively the "Duplex" type of machine, in which the two spindles are horizontal and opposed to one another, as indicated at **D**, could be employed—in the writer's opinion, to optimum effect.

When the component is of medium size and requires machining operations on both side and face, the "Duplex" machine again proves its value. It is seen being used to advantage at **E**, where one spindle is adjusted to mill the top face and the other the side.

The sketch at **F** in Fig. 5.2 shows a component having a face to be machined which varies in width, there being a space in the centre before machining is resumed at the far end. In work of this kind there is danger of time being lost because of:

(1) Table traversing during the blank;
(2) The operator having to wind the table across the blank; or
(3) The operator being unable conveniently to alter feed and speed to suit the reduced area.

To minimise these risks, a special type of machine has been introduced having a number of "dogs" clamped to the side of the table, which automatically trip the feed mechanism. In the case of the component at **F**, therefore, the table would traverse at a certain rate from *a* to *b*. The feed would then be tripped to a faster rate from *b* to *c*; from *c* to *d* it would move past at the full rate of, say, 30 m per minute, before being again tripped to the correct rate of feed at *d* as before. Finally, at *e* the table is tripped once more, running back to its starting place at the full 30 m per minute, when the component is removed.

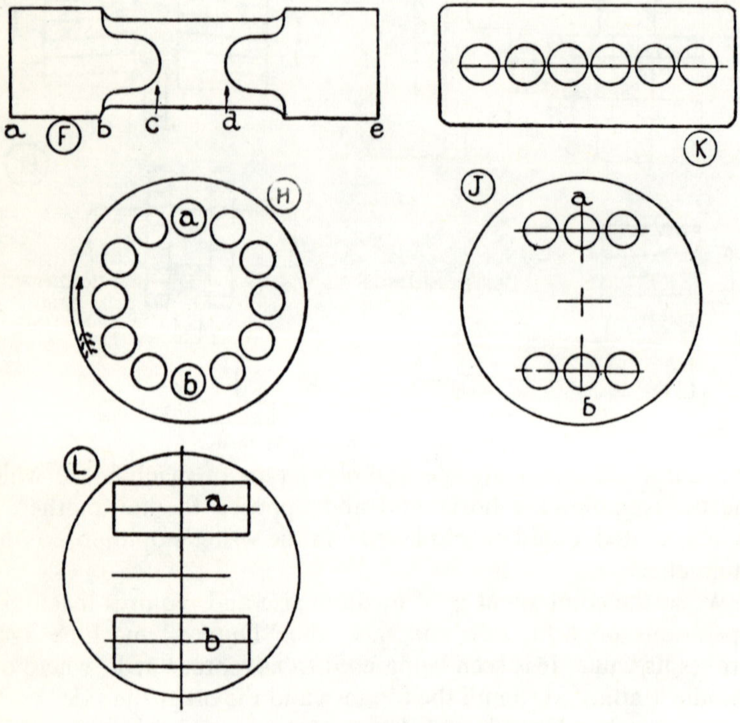

Fig. 5.2 More Milling Methods

For continuous work, the parts could be arranged on a fixture (**H** in Fig. 5.2) of circular form fixed to a rotary table used in conjunction with a vertical milling machine. The table is loaded and the revolving mechanism put in gear. While the cutter is at work at *a*, the operator is able to unload finished pieces at *b* and to replace them with raw ones.

Semi-continuous work can be carried out on the horizontal milling machine by means of the fixture shown at **J**, where the components *a* pass under the cutters while those at *b* are being loaded. When the milling of those at *a* is completed, the horizontal table is run back and the rotary table turned through 180°, so bringing the components at *b* into line for machining, while at the same time permitting the removal and replacement of those at *a*.

A more common method is to employ what is known as a "string" fixture. This is a fixture holding a number of components in a line, as at **K** in Fig. 5.2. Though the pieces can be removed as fast as they pass under the cutters, they cannot be replaced until the fixture has completed its traverse and has been returned to its starting point. This causes only a small waste of time, for the operator is able to load the first one or two pieces into the fixture and then set the machine in motion, loading the remainder as the cutting proceeds.

One other type of machine handles semi-continuous work. In it, the rotating main table, which is circular in form, carries on it at equal spacing from the centre two tables capable of longitudinal traverse. It is shown at **L** in Fig. 5.2. The parts are clamped in the fixture on the slide *a*, and while these are passing under the cutters the fixture *b*, which is identical to *a*, is being unloaded. Again, the loss of time when rotating from *a* to *b* is not serious.

5.3 Setting Spots. To facilitate the setting of the cutters relative to the fixture, setting spots are used. They vary according to the nature of the operation, but usually consist of hardened steel pieces permanently attached to the fixture in such a position that, when measurements are taken from them, the correct position of the cutter is ascertained without difficulty. A normal method is to decide on a standard amount of space to be allowed between the spot and the cutter for the latter to be correctly located. When no standard amount is fixed, the amount decided upon for a particular fixture must be plainly marked thereon for the information of the setter.

Fig. 5.3 shows how setting spots are used. Seen at **A** is a cutter milling the piece *b* to the height *d*, the latter being determined by placing between the cutter and the setting spot *e* a feeler having a thickness of exactly 0·25 mm so that guesswork is eliminated. **B** shows a slotting cutter which has to be correctly located both for depth and for lateral position. This is done by means of the setting spot *a*, which enables a feeler of the appropriate thickness to be inserted both below the cutter and at its sides.

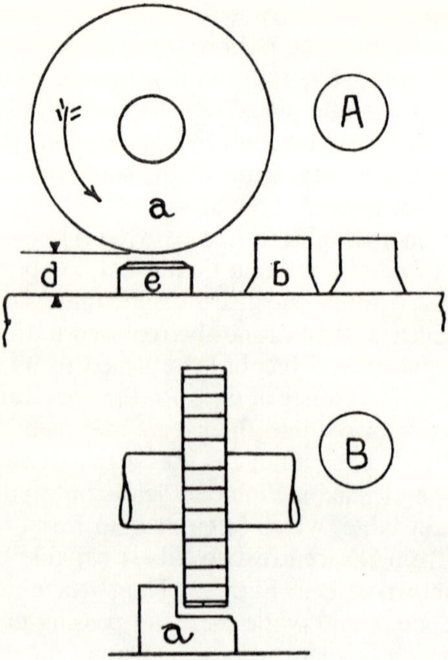

Fig. 5.3 The Use of Setting Spots

5.4 Designing a Simple Fixture. Fig. 5.4 shows a simple fixture designed to hold a single workpiece only—in this case an aluminium cover plate. It is shown in the Fig. shaded in position. The fixture is intended for use on a vertical-spindle milling machine—a type of machine in which there is always a tendency for the cutter to fling the component off the machine. The small fixing holes previously drilled round the edge of the component are therefore used to assist in holding. The mild steel pillar seen at *a* is surmounted by a small stud. The component rests on the pillar, and the stud enters one of the drilled holes to locate the piece before it is clamped. The piece requires machining on its face—which makes clamping a difficult problem since the cover plate is only 6·5 mm thick and made of aluminium. The difficulty is overcome in the manner indicated—but too much strain must not be put on the nut **H**.

The fixture consists of a cast-iron base **A** having four steel pillars as at **B**, and two pads **C** into which fits slide **J**. Nut **H** operates the sliding wedge **J** (seen better in the sectional elevation) to hold the component down. The component is dropped on to the fixture, and the knob **L** passes through the hole **K** in the wedge. Nut **H** is turned, causing the wedge to be drawn forward and at the same time drawing the narrow slot **M** of the wedge under the knob **L** so as to prevent the component from being lifted out. Since this may not always lock the component firmly, the under-side of the slot in the wedge is tapered slightly, as at **P**, so that as the wedge is drawn across it draws the knob **L** downwards, the latter being thus firmly held in position.

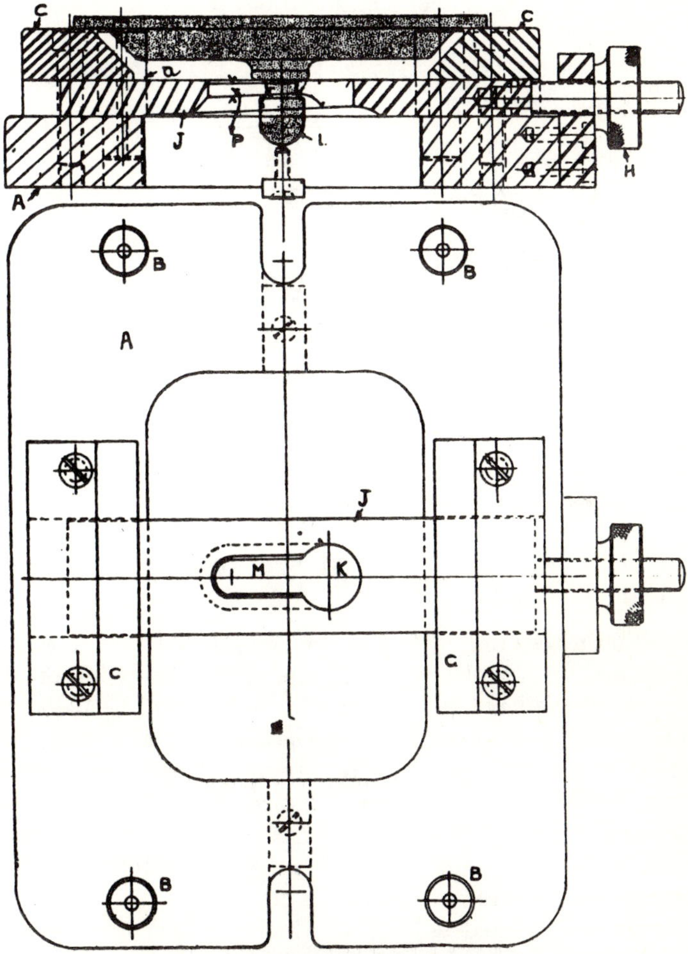

Fig. 5.4 A Simple Milling Fixture designed to hold an Aluminium Cover Plate

To prevent excessive leverage being applied to the knob **H**, it should be made with only a small head if it is knurled, or with a larger head if it is left plain. Any tommy-pins fitted to it should be left short. In a case like this, holes for tommy bars or flats for spanners should never be fitted lest an operator of low intelligence should insert a bar of excessive length (or, as is sometimes done, should fit over a short tommy-pin a length of steel tube). The object in both cases is to increase the leverage—but the result will be either to break the fixture or to spoil the work. It is always the duty of a designer to make every effort to render a fixture fool-proof.

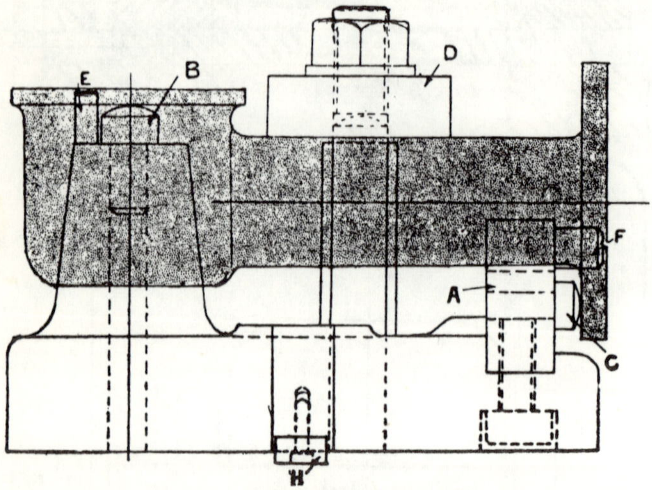

Fig. 5.5 A Milling Fixture for a Component to be Machined in Two Planes

Fig. 5.5 shows a milling fixture for a component which requires to be machined in two planes. The component is located in the fixture by means of the vee block **A** and the studs **B** and **C**. If location is effected under the flanges at **B** and **C**, their correct thickness will be assured; and if the operation is carried out on a duplex type of machine (see Fig. 5.1 at **E**) the two flanges will be machined at right angles. The clamp **D** is of standard pattern. The setting spots are located at **E** and **F**, and require a feeler-gauge 0·25 mm thick to ensure the cutter being in the correct position.

The tenon **H** is fitted with screws which can be changed to another size should it be necessary to transfer the fixture to another type of machine.

5.5 String Fixtures. As has been said earlier, this type of fixture occupies a considerable length of machine table and holds a number of components in line. Unless the machine is provided with a tripping mechanism enabling the table to pass rapidly across the blank spaces between these components, they must be carefully arranged so as to reduce blank space to a minimum.

The end view of a string fixture is, as it happens, well illustrated by Fig. 5.5, but the two views given in Fig. 5.6 are perhaps more specific.

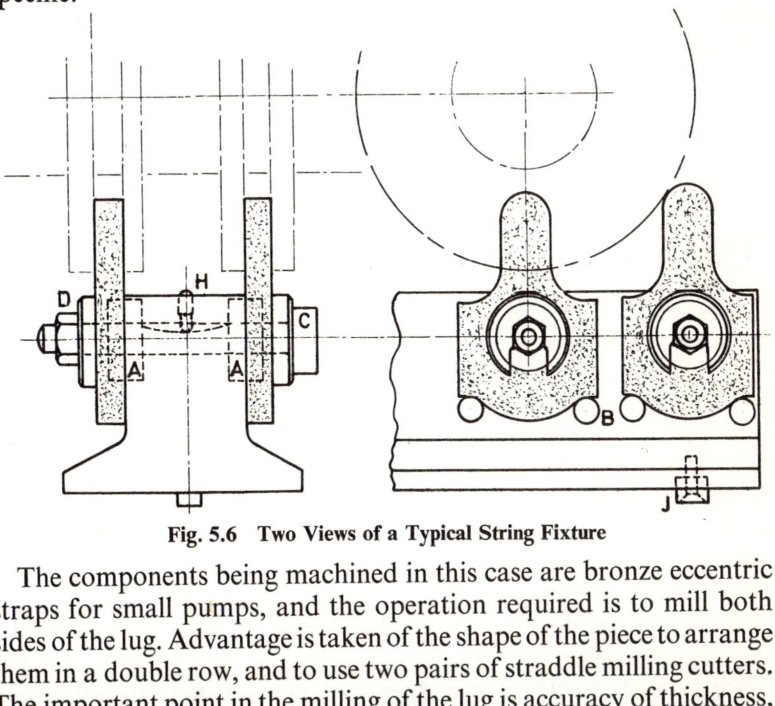

Fig. 5.6 Two Views of a Typical String Fixture

The components being machined in this case are bronze eccentric straps for small pumps, and the operation required is to mill both sides of the lug. Advantage is taken of the shape of the piece to arrange them in a double row, and to use two pairs of straddle milling cutters. The important point in the milling of the lug is accuracy of thickness, accuracy of depth being less important.

The component having already had its large hole bored, it is located by this on the two plugs **A**, with extra support given by the pegs **B**. To facilitate loading and unloading, the head of the bolt **C** and the clamp nut **D** are both made smaller than the bore of the component. The latter can thus be removed without taking off the nut. In operation, the components are slipped on to each end of the bolt, and "U" washers are placed behind them. When nut **D** is then tightened, the bolt **C** is drawn up and locks the pieces in position.

To prevent the bolt turning round when the nut is moved, a slot is milled along the length of **C**, and the grub screw **H** with the pin at its extremity is adjusted so as to allow the bolt to slide but not to revolve. The tenon **J** ensures that the fixture is set at right angles to the machine spindle.

The body of the fixture is of cast iron, broadened at the base to give rigidity. The pegs **B** are made of drill rod hardened and tempered, and the locating plugs **A** of medium carbon steel left hard. Bolt **C**, nut **D** and screw **H** are turned from case-hardened steel, the bolt being ground after hardening.

It is good practice when laying out most classes of fixture to indicate the type of milling cutters to be used, since this reduces the risk of unsuitable cutters being used in the machine shops, with a resulting loss of full productive value from the fixture.

Fig. 5.7 illustrates another useful design of milling fixture, this time for the rough milling of bearing caps for petrol engines before they are bored. The bolt holes are drilled in an earlier operation and are used for location—with only two of the four holes drilled in each piece being used for the purpose, one at either angle of a diagonally opposite corner.

Clamping of the fixture is effected by the two jaws **A** operated by the nut **B** on the bolt **C**. The latter passes through the nearest jaw, then through the body of the fixture and into the further jaw, the hole in which is about 2·5 mm larger than the diameter of the bolt. A small pin **D** passes through both bolt and jaw, thus providing a pivot. The bolt should ideally be positioned in the fixture as near as possible to the component being machined.

To prevent the jaws swinging out of position, they each have a hole which fits the guide pins at **E**, the hole being about 3 mm larger than the pins. The latter also have a shoulder on which the jaw itself bears. Since the outside of the components to be machined will still be in a rough state, the upper ends of the jaws are serrated or roughened to give them a more effective grip.

As in the string fixture illustrated in Fig. 5.6, the locking bolt is slotted and prevented from rotating by means of a screw **F**. The base or body of the fixture is of cast iron, the jaws of mild steel case-hardened, the pegs **H** of carbon steel hardened and tempered, and the bolt, nut and studs **E** of case-hardened steel.

Fig. 5.8 illustrates another type of string fixture, shown in operation during the milling of stub axles for automobiles. The noticeable rigidity of the fixture is especially necessary in this case, for the

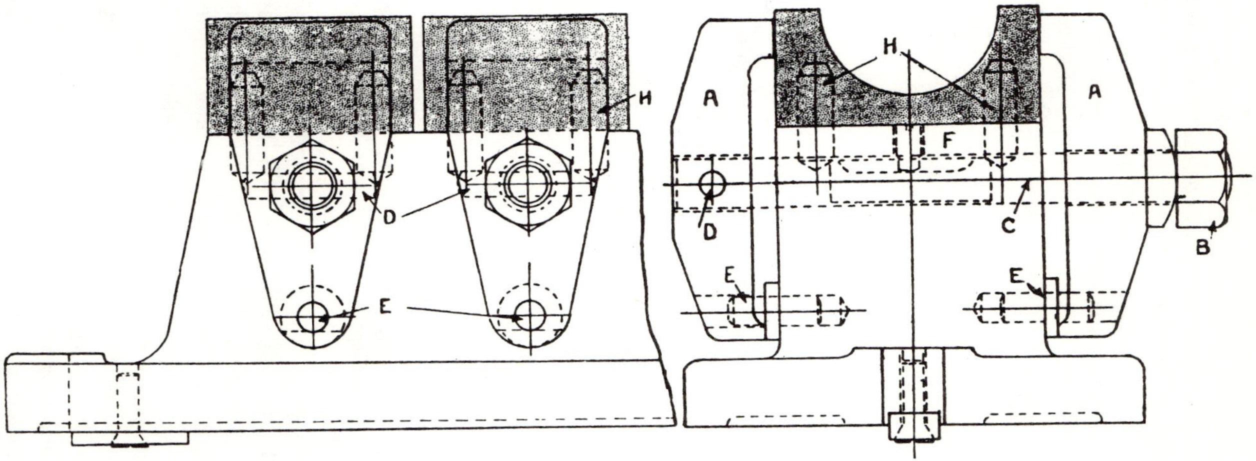

Fig. 5.7 A String Fixture used in the Rough Milling of Bearing Caps for Petrol Engines

components are of forged steel and the whole width of the gap is being machined in one cut. The cutter is of the "interlocking" pattern, being made in two pieces located side by side, with the teeth so cut as to overlap or interlock. This allows the two halves of the cutter to be forced somewhat further apart without leaving a gap between them when wear, or a resharpening, would otherwise cause the width of the milling to be narrower than it should be. The zigzag or "staggered" parting line can be seen in the illustration.

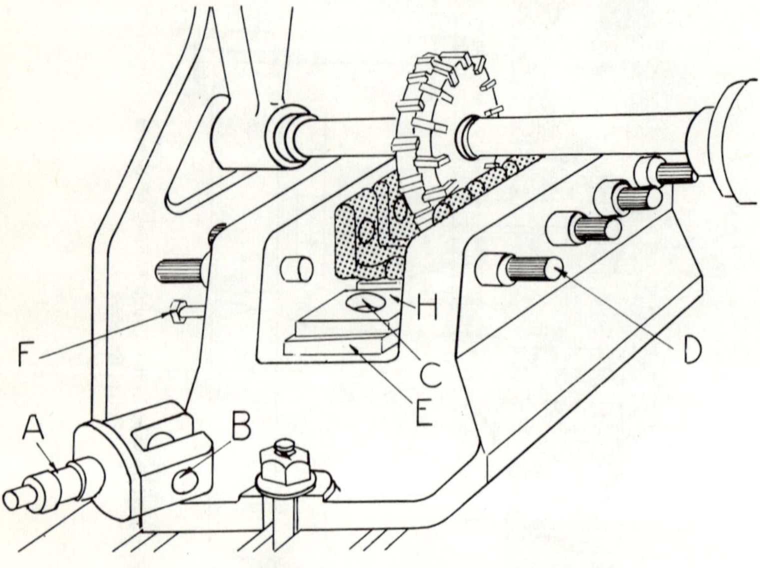

Fig. 5.8 A String Fixture with Interlocking Cutter used in the Milling of Automobile Stub Axles

Previous operations performed on the stubs had been to turn the stem seen at **A** and to drill the holes at **B**, and both are used for locating purposes. The stem of the stub is passed through the hole **C** and also through a slotted hole in the lower plate **E**. The horizontal plugs **D**, an accurate fit in steel liners forced into the main casting (the purpose of the liners being to prevent wear), are pushed from both sides into the holes **B** of the workpieces. The bolts **F** are then operated to draw sideways the plates **E**, which are of tapered section, so lifting plate **H** and firmly locking the work.

The hexagon-headed bolts **F** could perhaps with advantage have been replaced by another form not requiring the use of a spanner.

When the use of a spanner is unavoidable, however, one of either the tube or the box pattern will be found convenient.

5.6 Continuous Milling. Continuous milling cannot be carried out on a horizontal machine, but is quite simply done with a vertical spindle machine and a fixture located on a revolving table. Output by this method is generally limited by the operator's ability to unload, clean and reload the fixture while it is slowly revolving. Otherwise, save for the time taken in sharpening or adjusting the cutters, there need not be an unproductive moment.

Fig. 5.9 shows a portion of a circular fixture, in which the small levers indicated are to be machined on one face of the larger boss situated at their thicker end. It will be seen that the components are arranged radially round the fixture as close together as possible, with each pair being held by a single clamp consisting of a "U" washer

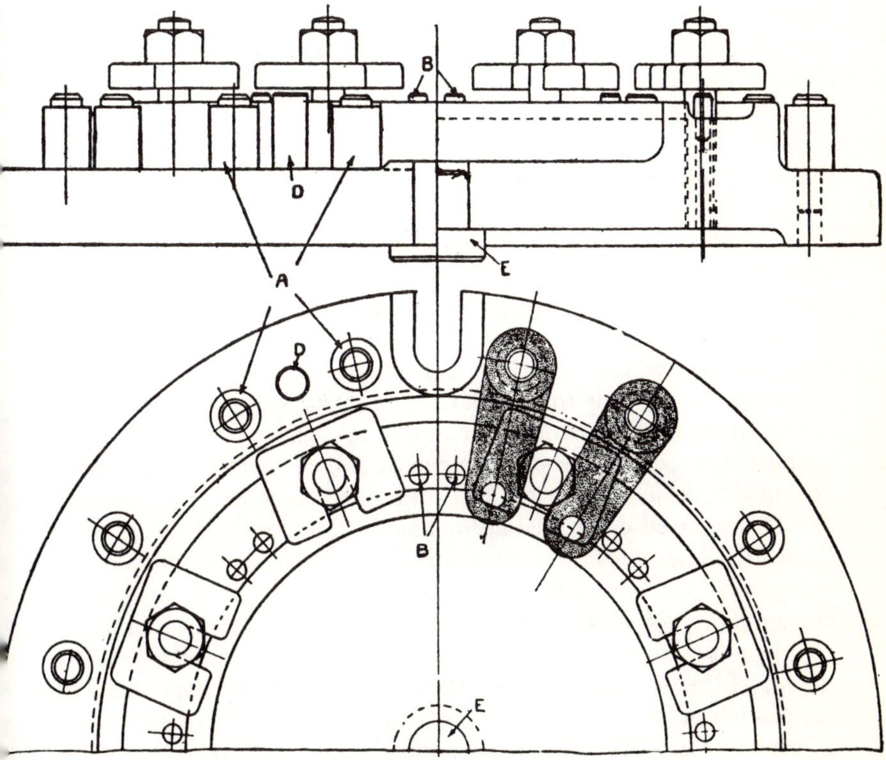

Fig. 5.9 A Circular Fixture used in Continuous Milling

held by a nut. A hole previously drilled in the centre of the boss to be machined makes a suitable locating point, and for this purpose the shouldered pins **A** are arranged round the fixture. The opposite ends of the pieces are kept in position by the pegs **B**.

The setting spot is placed at **D**, and it is not unusual to add one or two more of these equally spaced round the fixture for the convenience of the setter. The steel peg **E** in the centre of the fixture is a push fit into a hole in the centre of the revolving table whose purpose is to ensure that the fixture is centrally located thereon.

The body of the table is of cast iron, the various pins of carbon steel hardened and tempered, and the washers, etc., of case-hardened mild steel.

A fixture of this pattern could also be used on a horizontal machine. Though it could not revolve continuously, a portion of one of its halves, already loaded, could be passed under the cutter while the opposite portion of the fixture was being loaded. When the first loading had been milled, the table would be run back and the fixture rotated through 180°, so bringing the next batch into position in front of the cutters. The cut would then be started while the last batch was being unloaded, and so on.

5.7 Indexing Milling Fixtures. It is often necessary to repeat an operation on another portion of the same component while ensuring that the relative position of the two faces, etc., machined shall remain absolutely constant. Fig. 5.10 illustrates a fixture embodying a mechanism designed to achieve this purpose. The component, a hexagonal iron casting, requires machining on six sides, having been previously bored and faced. The operation is performed on a duplex machine, so that only three indexings are necessary.

Location of the casting is achieved by means of an expanding mandrel comprising the three segments **A** held together by the steel spring ring **B**. The segments rest on a cone piece **C**; and when the nut **D** is screwed down, the upper cone **E** descends and forces them outwards. The vee-shaped piece **F** is slackened when the faces of the component are being arranged square on loading, but it engages the corner of the casting when the expansion described above takes place.

When the component is being rotated, the vee is held back by means of the peg **H**, which is drawn back and given a half-turn so as to catch on the boss at **J**. The wide base of the lower cone **C** acts as a platform on which the component is clamped so as to revolve with it. The outer edge of the platform has three slots **K**, milled 120°

apart, into which drops the lever **L** under the influence of the spring plunger **M**. The three studs **P** form a seating for the component.

The base of the fixture is of cast iron and the remaining parts, except the springs, of case-hardened steel.

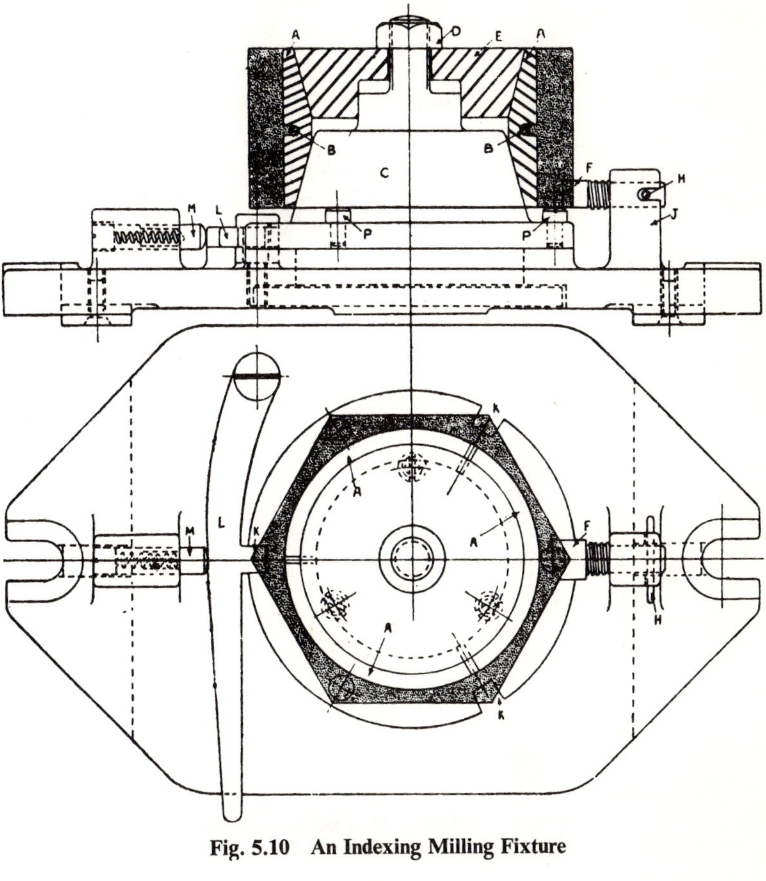

Fig. 5.10 An Indexing Milling Fixture

CHAPTER 6

Mandrels, Chucks and Turning Equipment

6.1 Despite the constant search for methods of quicker and cheaper production and the consequent tendency towards early elimination of old machinery and methods, the *centre lathe* in various improved forms is still in widespread use, and there are some kinds of work which can be successfully carried out only on this machine. It is therefore important to consider methods of handling work on the lathe, beginning with the one most commonly employed.

6.2 Mandrels. The operation of a mandrel cannot be entrusted to a man who is less than semi-skilled. Nevertheless, their construction should be made as simple as possible.

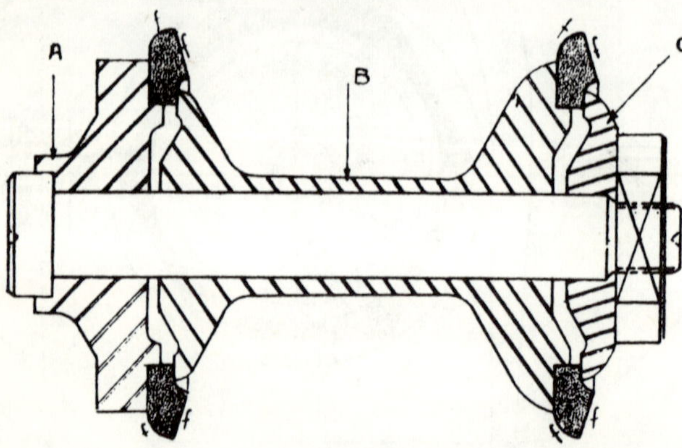

Fig. 6.1 A Mandrel used for Multiple Turning on an Automatic Lathe

Fig. 6.1 illustrates a mandrel for use on a "Fay" automatic lathe. The components to be machined are shown shaded in position. They consist of steel ring bevel gears, the operation required on them being to machine the rim and face indicated—the former by means of a tool of dovetail form and the latter with ordinary turning tools. Since the gears in question were some 150 mm diameter, the rather large mandrel was made in several pieces, the parts **A**, **B** and **C** being steel forgings. The mandrel itself should be of case-hardened steel, hardened and then ground.

On the particular lathe chosen, which happened to be arranged for multiple turning, two components could be machined at once. It was therefore necessary to maintain very carefully the correct positional relationship on the mandrel between the two components.

For a casting which is long in proportion to its diameter and which is to be turned from its cored bore, a cheap form of centering mandrel is shown in Fig. 6.2. The component in this case is an automobile axle sleeve; the rough-cored hole through this sleeve must be made reasonably true with the outside diameter.

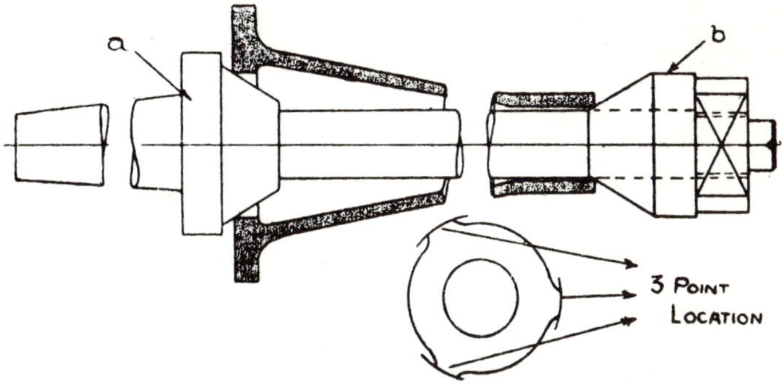

Fig. 6.2 A Mandrel used in the Turning of an Axle Sleeve whose Outside Diameter must be Concentric with its Cored Bore

The coned boss *a* is integral with the shaft, which is tapered to fit the nose of the machine. The coned piece *b* has the same cone angle as *a* (60°), but is free to slide on the mandrel. Since irregularities of the ends of the bore could cause difficulties, the two cones are relieved, leaving three points to engage the casting. The fact that at least one of the cones is capable of revolving is of great help in getting the component to run concentric.

Automobile work calls for the turning of a large quantity of gears having a splined hole of the form shown at *a* in Fig. 6.3. Since these are broached out, they can (given reasonable care) form a reliable surface to work from.

The end view of the same Fig. shows a section of the splined mandrel with the gear in position. Note that at *a* the *top* of the spline of the gear seats on the *bottom* of the spline of the mandrel. This is purposely exaggerated on the drawing, since it is desired to emphasise the point that the chucking attachment used should always be as near as possible to the final working position of the component.

Moreover, though it might be possible to use a good round mandrel having no spline, the presence of the latter not only acts as a positive drive during turning but also virtually assures final concentricity of the gear.

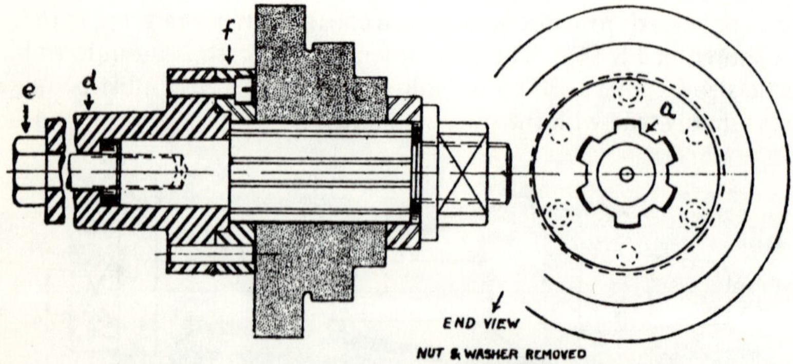

Fig. 6.3 A Mandrel used in the Turning of Splined Gear Blanks

On the mandrel shown, the tapered part *d* is so arranged with the draw-bolt *e* that it is possible to fit other mandrels of different sizes. The pad *f* is made of hardened steel in order to minimise the wear which would cause the component blanks to run eccentrically.

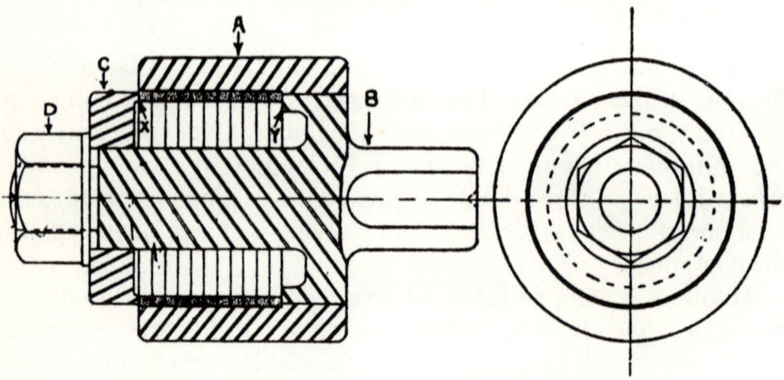

Fig. 6.4 A Mandrel used in Grinding the Outside Diameters of Split Piston Rings

The mandrel illustrated in Fig. 6.4 is intended for grinding the outside diameter of piston-rings, a number of which are shown in position, having been previously split. But the principle on which it works can also be adapted for turned work, if desired. The fixture is used in the following way. The rings, which must be clean, are placed together in the sleeve **A**, a little pressure being needed as the slots must be compressed to get them in. The filled sleeve is slipped on to the mandrel proper **B**, and the washer **C** is put on its opposite end. When the nut **D** is tightened up, the rings are effectively clamped between the rims **X** and **Y**. The sleeve **A** is then removed.

Note that sleeve **A** must have a long bearing surface and must be a sliding fit on the mandrel without any slackness, lest the rings be not loaded concentrically. Washer **C**, also, must fit the mandrel accurately. Its outside diameter should be as large as possible after allowance has been made for the amount of metal needing to be removed from the rings.

6.3 Expanding Mandrels. Expanding mandrels are of especial use where the work is "second operation" or rechucking work—that is to say, when a previous operation has produced a machined surface sufficiently accurate for locating to be done from it. Such mandrels are not, however, suitable for heavy turning in cases where the amount of metal to be removed is great or the diameter of the component is large in proportion to its width. The reason is that the absence of a positive drive often allows the component to turn on the mandrel and to become loose as soon as the tools commence to cut.

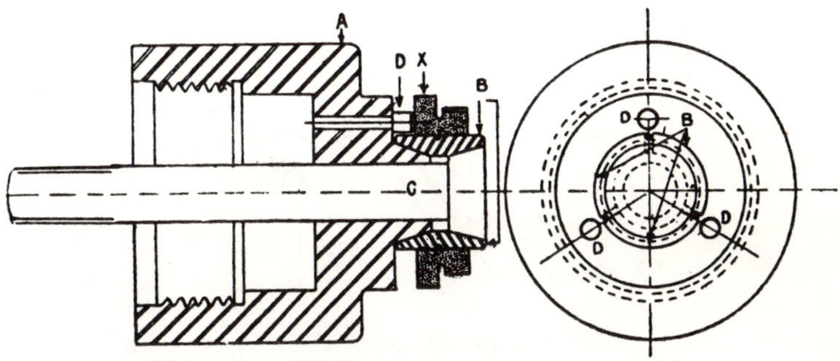

Fig. 6.5 An Expanding Mandrel making use of the Drawback Motion of the Capstan

Fig. 6.5 illustrates a type of mandrel which has proved successful. The hood or nose-piece **A**, made of mild steel carbonised and hardened, fits the screwed nose of a capstan lathe. The diameter of the three steel segments **B** is dictated by the diameter of the component *x* being machined—in this case a piece of switch-gear. The segments are made of hardened steel to prevent wear; and they slide up the coned portion of **A** when actuated by the coned bolt **C**. This latter is moved by a rod passing through the machine spindle and operated by a hand-wheel, or else the usual clutch toggles fitted to the machine are adapted to give the necessary pulling motion.

To keep the segments on the cones, they are fitted with a groove in which rides a round steel spring. This spring, while holding the segments in position, allows them to expand when necessary. To assist the component to settle itself squarely on the mandrel so as to give it a positive seating from which to work, three hardened steel studs **D** are fitted symmetrically at a suitable pitch circle diameter. The component is held firmly up against them with one hand, while with the other the operator activates the mechanism which tightens the coned bolt.

The angle of the cones should not be too great, about 30° being considered normal. The nose-piece **A** can be made of cast-iron where the diameter is over about 125 mm, but the coned portion must then be fitted separately and must be made of hardened steel.

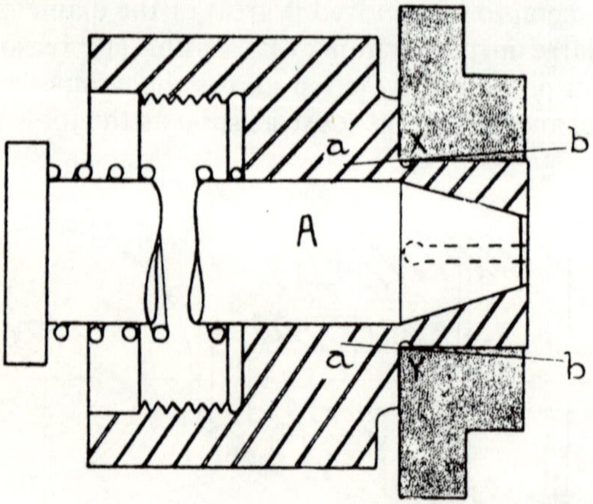

Fig. 6.6 An Expanding Mandrel making use of the Push-out Motion of the Capstan

Another form of expanding mandrel utilising the ordinary "push-out" motion of the capstan is illustrated in Fig. 6.6. Note that the expanding portion and the nose-piece are now combined, the expansion being effected when plunger **A** is forced into the coned nose. Note also that the outside diameter of the nose does not expand equally along its length, as it did in Fig. 6.5, but pivots at **X** and **Y**, thus causing the front to be of larger diameter than the back. This is indicated by the lines *a*, *b*. If the bore of the component is reasonably parallel and accurate to size, this will not be sufficient to cause trouble; but if the bore varies appreciably, the component will be held only on its edge, the grip will be reduced, and the piece will work loose immediately the tool touches it. A positive drive can be obtained if the component is given a hole through its face and the nose-piece of the mandrel is fitted with a steel peg.

The writer remembers the case of a soft brass tube, 20 mm long, 30 mm bore and about 1·5 mm thick, which had to be threaded on the outside at about 1 mm pitch. When this was being done, some 70 per cent of the tubing was pulled off the mandrel when the die reversed, and had to be chipped out of the die with a chisel. When a self-opening die was used, other bits twisted on the mandrel after a few threads, the rings appearing to expand with the mandrel so that they could not be gripped. The trouble was overcome in a simple way. The customer's permission was obtained for a rather heavy centre punch-mark to be made on the outside of the rings, such as would not interfere with their final use. When the rings were put on the mandrel, however, the bulge or "pimple" caused to appear on the inside of the rings by the centre punch-mark made it impossible for the rings to be slid on to the mandrel unless all the "pimples" engaged in one of its slots. When the latter was expanded it now gave a good positive drive, and the outside of all the rings was threaded quite easily.

6.4 A Special-Purpose Expanding Mandrel. It has already been said that components with cored holes are difficult to handle on expanding mandrels, but it is sometimes essential that this method of holding such components be used. When this is so, a special mandrel must be designed. Fig. 6.7 shows one designed to hold cast-iron pistons while they are being turned.

Since it is essential that the outside of the component be kept reasonably concentric with the bore, which is rough as cast, some form of gripping device that will work by being applied to the inside walls is necessary. The thick parts of the component at **B**, where the grooves will be cut, are inconveniently placed for the purpose because they are some distance apart, so the ordinary methods of cone expansion cannot be used. Fig. 6.7 shows how the problem can be solved, with **A** the main body of the fixture internally screwed to engage the nose of the lathe. Six holes are drilled radially in the nose portion of the body, equidistant round its circumference. Through this circumference protrude the gripping pegs **C** and **C_1** (two only being shown to avoid cluttering up the drawing).

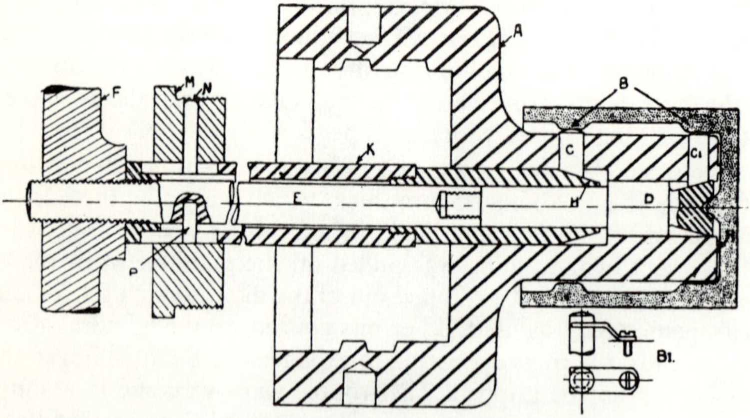

Fig. 6.7 A Special Expanding Mandrel used to hold Cast-Iron Pistons during Turning

The front pegs **C_1** are expanded when the coned bolt **D** is drawn back by means of the rod **E** and the wheel **F**. The back pegs **C** are by the same movement forced out by the cone **H**, which is connected to the sleeve **K**. Thus when wheel **F** is turned, cone **H** is pushed forward and the cone at **D** is drawn in until the pegs grip the casting. The boss **M** is fixed at the rear end of the spindle to support and centralise the rod **E** and the sleeve **K**. To prevent either of the latter turning when the mechanism is operated yet at the same time to allow end motion, the sleeve **K** is fitted with a retaining pin **N** and the rod **E** with a pin at **P**.

To assist the six pegs **C**, **C_1**, etc., to return when the wheel **F** is released, each is provided with a spring (shown only in the detail drawing at **B_1**) which is of just sufficient strength to cause them to return far enough to clear the component as it is drawn off.

The piston butts up against the three small projections **R** at the end of the nose **A**, thus ensuring that the piston tops are of uniform thickness. To add strength to the rather weak nose part of **A**, the radius at its junction with the main body should be made as large as possible; and the diameter of the nose should also be made as full as possible, with only the radius of the pegs showing above it.

In another type of special-purpose expanding mandrel, the pegs are replaced by two segments which expand on the double-cone principle described above; but with this type loading and unloading become difficult if the inside surfaces of the castings vary in form or are noticeably irregular.

6.5 Spring Collets. These devices find employment in a number of different machines, and vary in form from the simple example illustrated in Fig. 4.10 to the more elaborate types fitted with adapter pads for use on automatic lathes. The normal type of collet used on capstan lathes needs no description here, but a special type is illustrated in Fig. 6.8. It is called an "extra-capacity" collet, because it will hold a component having a diameter larger than the size of the spindle bore (*i.e.*, the "capacity") of the machine.

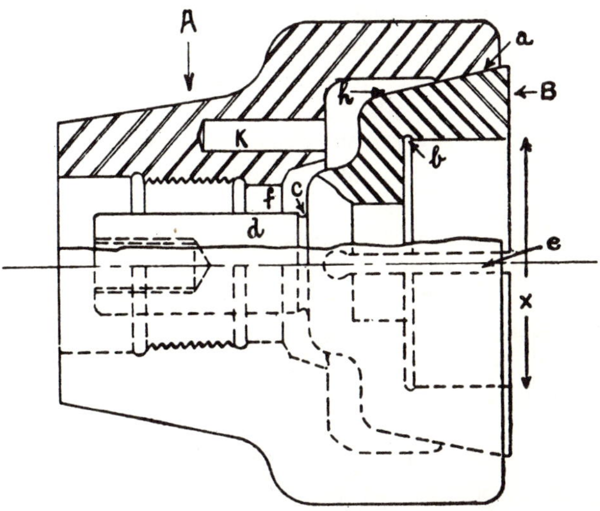

Fig. 6.8 An "Extra-Capacity" Spring Collet able to hold a Component whose diameter is larger than that of the Spindle Bore of the Machine

The device is seen to consist of a nose-piece **A** made of mild steel, which is case-hardened and ground on the coned portion *a*. Its other end is threaded to enable it to be screwed on to a particular machine. The collet **B** itself is usually made of carbon steel and hardened, the rear portion being finally tempered to give the necessary properties of spring. Mild steel case-hardened has also given fair results, but it does not retain the springiness necessary for free working.

Collets of this type are expensive to produce, so the draughtsman must try to avoid thin sections and sharp corners in the design. The presence of these features has caused the spoiling of many collets in the hardening process, or their failure in subsequent use.

The thickness of the shell of the collet must be made as even as possible, with all sharp corners avoided. Where one is unavoidable, it is best to put in a small radius, as at *b*. The small radius at *c* is there to enable the grinding wheel to work right up to the back of the collet when it is grinding the stem *d*. The latter must be a good sliding fit in the machine spindle. The slots *e* give the necessary spring. In this case they are four in number, equally spaced round the collet and extending to its rear face. A hole is drilled at the closed end of every slot to relieve the strain and so avoid fracture when the collet is contracted.

To ensure the true running of a chucking mechanism of this type, the draughtsman should whenever possible ask that the principal diameters be ground when they are in position on the machine. In other words, after the nose has been threaded and hardened, it should be screwed in position on to the machine which is to be used, and a portable emery grinder should be employed to grind the cone *a* perfectly concentric; for it is on this part that the collet depends to ensure that it sets true. When the surfaces *h* and *d* have been ground, the collet can be fitted to the nose-piece by whatever "drawback" mechanism it is intended to employ; and the diameter *x* is then ground to ensure perfect concentricity.

To prevent the collets closing or opening out at any time during manufacture (it generally occurs when they are heated and quenched to harden them), the slots are not allowed to break completely through the front face of the collet. Instead, a thin section of about 3 mm of metal is left so that the collet retains its shape while being hardened and ground. The section is finally removed with the aid of a thin elastic emery-wheel.

One difficulty with this type of chuck is that a slight variation in the diameter of the component will prevent the collet from drawing back far enough in its seating. When the operation to be performed on the piece is turning or grinding to a certain length or thickness, this variation in diameter will cause a variation in thickness also. One way of overcoming this is to insert into the holes placed as at **K** steel pegs which can protrude either through the slots in the collet or through holes specially drilled in them so as to provide a positive seating for the component.

It is also possible to place in the hole **K** or on the pegs used a helical spring which will help the collet to overcome any tendency to stick when it is released.

6.6 Thread Chucks. It is sometimes necessary to perform on a threaded component certain further operations which must be perfectly concentric with its threaded portion. When this is the case, the threaded portion of the component itself must be used for purposes of gripping—a practice not generally to be recommended but sometimes unavoidable. If it is carried out carefully, no trouble should result.

Fig. 6.9 illustrates two methods of holding a component by a previously screwed part. The upper one of the two shows how use can be made of the drawback action of the capstans, and the lower one of the "push-out" motion.

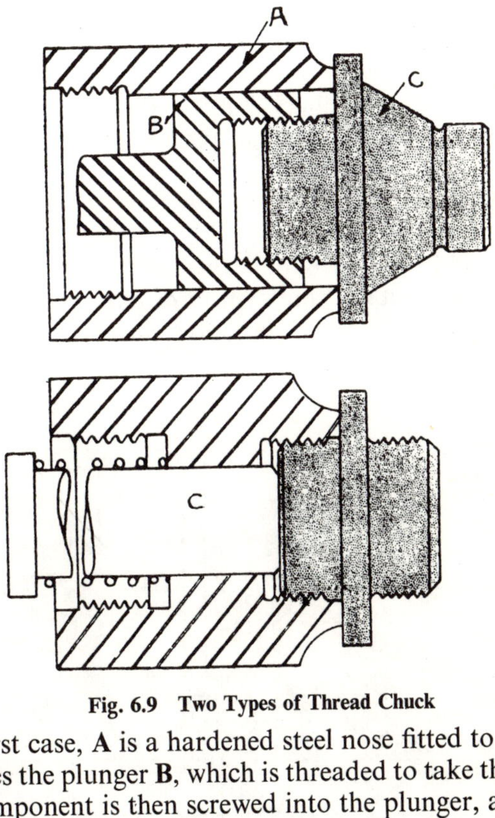

Fig. 6.9 Two Types of Thread Chuck

In the first case, **A** is a hardened steel nose fitted to the machine. Into it slides the plunger **B**, which is threaded to take the component **C**. This component is then screwed into the plunger, and the chuck lever of the machine is released until **C** butts against the face of the nose **A**. When the chuck lever is closed, **B** is drawn back and the component is effectively locked in position.

To remove the component, the operations described are performed in the reverse order.

In the type shown in the lower half of Fig. 6.9, the same principle is applied save that the pad **C** is thrust against the component, so giving the latter a seating which helps to take up some of the cutting thrust. Though simpler to arrange, however, this configuration is one to avoid, for it is always bad practice to thrust the component away from its seating.

In both cases, it will be seen that if the component had been screwed into a plain nose without any pads, the tightening effect of the cutting tools in the machining operation would have made the release of the component difficult. The fit of the component into the nose or pad, however, must in both cases be good lest the threads get distorted in the jambing action; and the adjustment of the clutch on the machine must not be too heavy.

Both types of threaded chuck illustrated can be adapted to take components having female threaded parts.

6.7 Turning Fixtures. Many pieces of awkward shape require turning operations, and a great variety of special holding devices have been developed to handle them economically—the object, as always, being to save both time and the employment of unnecessarily skilled and expensive labour. Fig. 6.10, for instance, illustrates a turning fixture

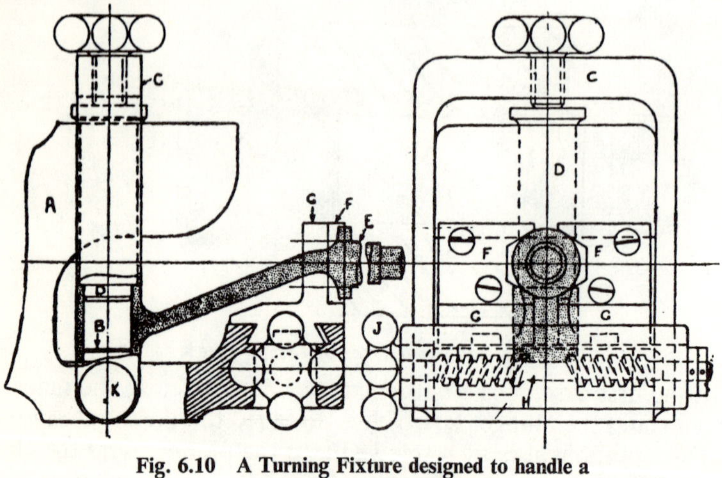

Fig. 6.10 A Turning Fixture designed to handle a Forging of Awkward Shape

designed to handle a steel forging (indicated in position in the Fig.) of remarkably unhandy outline.

The fixture was designed for use on a capstan lathe, the body **A** being screwed to fit into the machine nose. The boss of the workpiece had previously been milled on both sides, and a hole had been drilled through it which provided a good locating point by seating on the stud **B**. It was finally locked on to **B** by means of the swing clamp **C**, which holds the plunger **D** into the other side of the boss.

To centralise the round stem of the component **E**, some form of self-centering clamp was necessary, and this was achieved by fixing two vee-pieces **F-F** to the slides **G-G**. These, in turn, are actuated by the screw **H**, which has a left-handed thread at one end and a right-handed thread at the other. When the knob **J** is turned, the two slides are drawn together and centralise the stem.

The clamp **C**, swinging on its stud **K**, not only partly holds the component but also acts as a binding unit on the chuck, which is necessarily of peculiar shape. The clamp screw and the mass of metal about it serves in addition as a counter-weight to the other half of the chuck casting.

6.8 Indexing Fixtures. Components requiring a number of holes in the same plane can be finished on either a boring machine, a drilling machine or a capstan lathe. For a piece of exceptional weight and large dimensions, a boring machine will generally be best; but for one of rather small size, a good drilling jig will normally serve. It is when the work is of medium size and weight, and especially when the bores required call for the use of tools of different kinds (boring, recessing, threading, etc.), that the capstan lathe comes into its own.

An example of this type of machine is shown in Fig. 6.11, where the crank-case of a radial engine is seen being machined—the operation required being to rough-bore, finish-bore, face and screw nine holes to a limit of 0·025 mm. The job was performed in four hours. The cast-iron body **A** is screwed to the machine nose in the usual way, the mass of metal at the top being added to balance the fixture during operation. The workpiece is located on a platform **B**, and is held there by three plain clamps locked with the aid of a box spanner. The platform is fitted with nine index holes—one for every hole to be bored. An index plunger of the pattern shown in Fig. 4.11 would be satisfactory for the job—a pin of tapered form finding its location easier and automatically adjusting itself to the hole so that slackness and consequent rocking of the work is avoided.

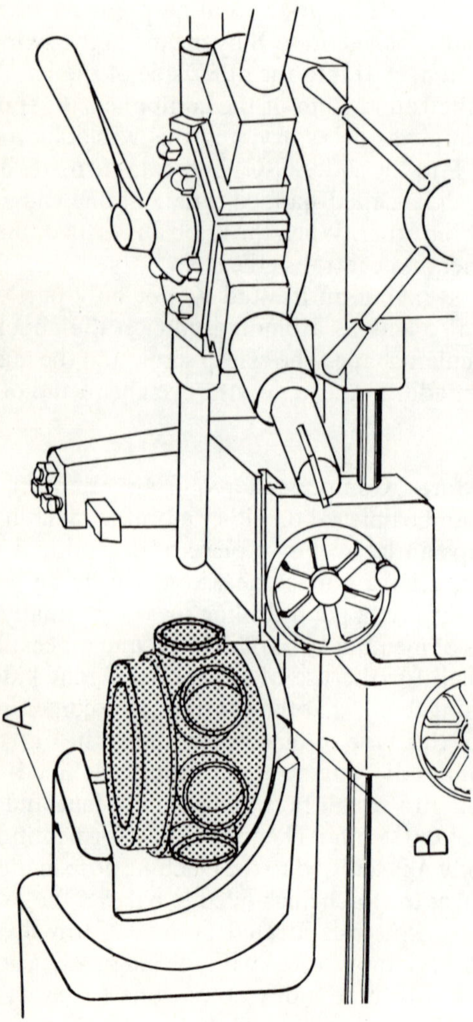

Fig. 6.11 An Indexing Fixture used for Machining the Bores of a Radial Engine Crank Case

Take care with all revolving fixtures to see that any plungers, clamps, etc., are so arranged that centrifugal force will not tend to loosen them (which would be very dangerous). Remember, nevertheless, that the value of the fixture can be lost if the indexing and locking mechanisms are made too complicated.

Another form of indexing fixture depends for its operation on a lateral movement instead of a revolving one. An instance would be the boring of a twin-cylinder motor casting or of the two bores of a water-meter. The slide carrying the work in either of these cases would be operated by a rack and pinion, and would be indexed for each hole by the usual plunger. Means would be provided to lock the slide before the fixture is allowed to revolve.

6.9 Cross-Slide Tool-Holders. It is not often necessary to design tool posts to hold ordinary turning tools of the flat type, because they are normally supplied with the machine. But since those in general use vary greatly in their degree of usefulness, a few words on them may be of value.

The first essential in a tool post is that, once the tool has been accurately set, it should hold the tool in that position in all normal conditions. It should also be easily clamped, and should be provided with some form of fine adjustment.

A comparison of the two types of tool-post illustrated in Fig. 6.12 throws up some interesting points. In type **A**, the tool *d* passes through the column *c* and rests across the top of nut *a*. This screws into the lower nut *b*, so that it becomes possible to adjust the tool either upwards or downwards.

A poor feature of this design, however, is that the tool is held by one screw only, which allows it to pivot when under cutting load. Furthermore, the clamping screw *e* is situated directly over the hole through which the stem *c* passes. The result will be that if the tool happens to be somewhat weak in the shank, or else of thinner section than is normally used with the tool-post in question, it will tend to bend when screw *e* is tightened. The line of bend is shown, exaggerated, by the line **XY** in Fig. 6.12.

Looking at the same tool-post in view **C**, it will also be seen that it is impossible to set the tool close to the machine nose without excessive overhang.

A generally superior design is seen at **B**, where the body *h*, with its base broader than its top, gives rigidity and the tool, held by the two screws *ee*, cannot swivel. Adjustment is achieved by the screw *b* engaging the wedge *a*, a turn of the screw in the appropriate direction either raising or lowering the tool *d*. A further advantage of this design is that the tool lies flush with the side of the tool-post and so can be brought within 1 mm of the spindle nose. This is shown in the right-hand illustration of view **C**.

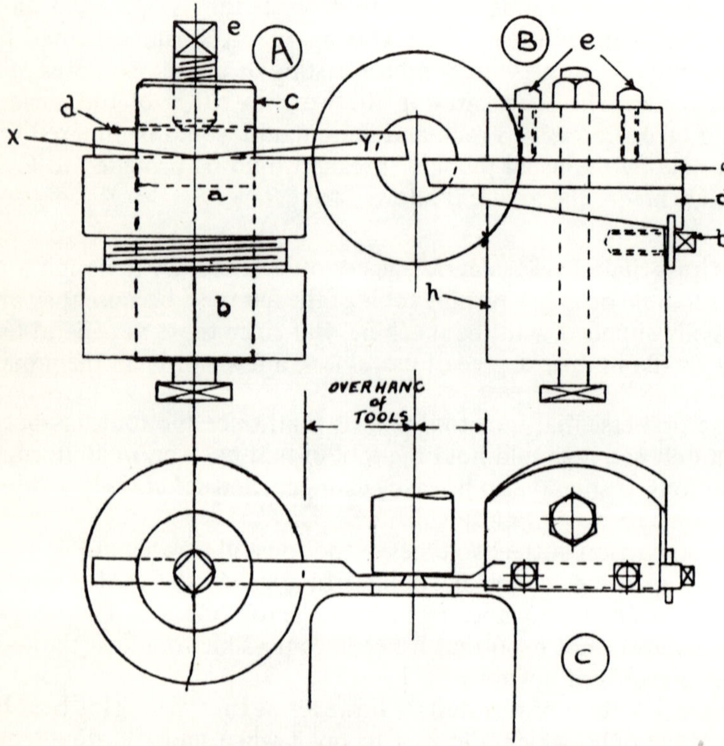

Fig. 6.12 Two Different Designs of Tool Post

6.10 Holders for Tools of Circular or Dovetail Form. Tools of this type require special holders, for unless they are correctly held trouble will always result. A means of fine adjustment is also essential, especially when the tools are of circular form.

Of the numerous patterns of these holders in existence, a useful pair are shown in Fig. 6.13. The circular tool-holder consists of a cast-iron bracket **A** provided with slots **B** for holding-down purposes, and with a tenon **C** fiittng into the slot in the machine cross-slide so as to ensure that the bracket sets at right angles to the machine spindle. The bolt **D** also passes through the bracket, and its head is sunk into the tool so as to enable the latter to be used close up to the machine nose. The bolt is keyed as at **F** so as to prevent it turning round when tightened.

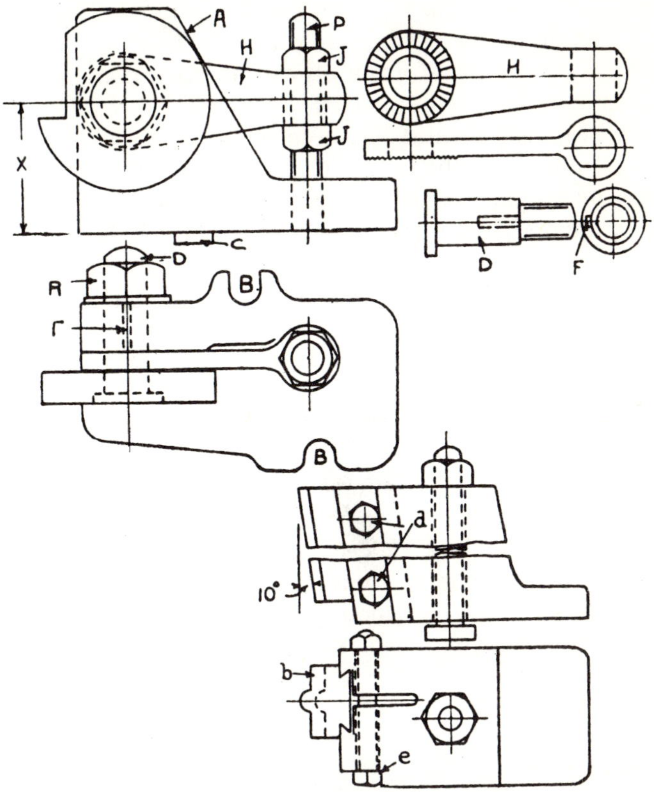

Fig. 6.13 Holders for Tools of Circular and Dovetail Form

Fine adjustment is obtained by means of the plate **H** and the nuts **J-J**. The detail drawing of plate **H** shows that one side is perfectly flat, while the other has a serration. At the tail of the plate is a boss through which is drilled a hole. Stem **P** clears through this hole, and the boss is secured on this stem between two nuts **J-J**. The circular tool itself is provided with serrations matching those on the plate so that, when it is assembled with the latter, the two serrated faces mesh and form a secure lock. To make a fine adjustment, it is only necessary to loosen the binding nut **R** and then raise or lower plate **H** by operating the nuts **J-J**.

The serrations mentioned are sometimes (but not very satisfactorily) replaced by equally-spaced holes in the plate which are engaged by a peg fixed in the side of the tool. Fine adjustment can also be secured by replacing stem **P** and the nuts **J-J** by a lever and nut. A cam on the latter raises or lowers the lever.

Holders for dovetail-form tools can be made either of mild steel or of cast iron. The angle *a* (see lower part of Fig. 6.13) is either 10° or any other angle which has been made standard in a particular works. The body of the holder is slotted to enable the tool *b* to be gripped by a tightening of the bolts *d*. To prevent these bolts from turning, the body is cut away at *e* to engage the bolt head.

A really good method of obtaining fine adjustment has yet to be devised; but it is sometimes achieved by putting a short bolt fitted with a nut under the tool so that when the nut is unscrewed the tool is forced up. This method has the further advantage of acting as a positive stop if the tool should develop a tendency to work down the holder, either because it is taking an excessive cut or perhaps because the fit of the dovetail is bad.

This dovetail should, of course, always be of a standard form.

6.11 End-Working Tool Holders. Holders for end-working tools—*i.e.*, drills, taps, counter-bores, etc.,—are customarily purchased from the builders of the machine tool, but the writer holds that it is cheaper to make even a small quantity of them oneself provided that the points really calling for attention are watched and the non-essentials neglected.

6.12 Plain Holders. The simplest possible type of holder is shown in the upper diagram in Fig. 6.14. It works quite satisfactorily provided it is kept in good condition. The body **A** is made of case-hardened steel, and is ground on the stem and in the bore as indicated. It is a waste of time to add a fine finish anywhere else. The stop-screw **B**, though often omitted, is of great value, for it prevents any possibility of the tool being pushed back. When short ends of tools are being used up, it is easy to insert a piece of steel of suitable length to fill up the space.

A small but important point concerns the end of the screw **D**. This should have two or three of its end threads removed and should then be hardened to prevent it spreading or flattening as a result of continual tightening-up. Its front bore must be ground exactly concentric with the stem.

The collet type of end-working tool-holder shown in the lower part of Fig. 6.14 is much used to hold drills of small size, from 10 mm downwards. Note, however, that with this method of gripping, the stop-screw indicated at **B** should never be omitted. The body of the holder should be hardened, and both the body and the coned mouth ground so that both shall be concentric and well finished. The cap **A** is made of the same material as the body and, since it is subject to hard wear, should also be hardened.

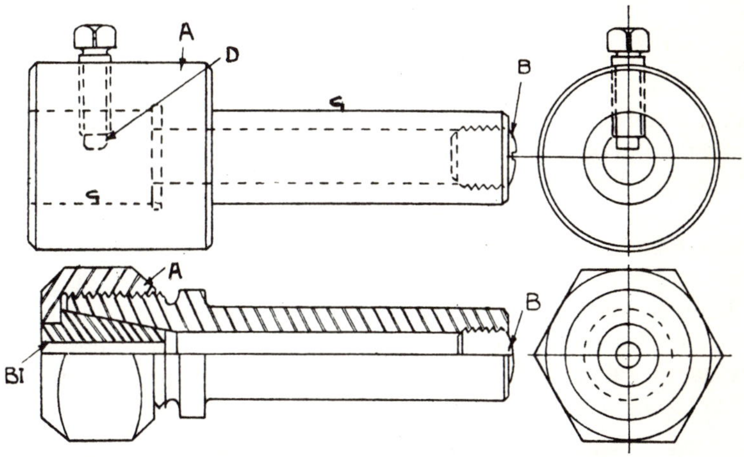

Fig. 6.14 A Plain and a Collet-type End-working Tool Holder

The collet B_1 should be made as long as possible, so as to give a good bearing surface and a better grip. The fairly common practice of splitting it into separate segments, however, is one to be avoided—a warning which will not be necessary if the reader has ever had to search for any small part of a tool or tool-holder which has dropped into the oil and swarf collected in the machine bed.

6.13 Bushes for Tool-holders. These unimportant-looking little pieces play a large part in the making of a successful tool-holder. The one shown at **A** in Fig. 6.15 is split through one side and partly split through the other. The bottom of the slot is fashioned in the form of a radius to prevent the bush breaking.

The bush shown at **B** has two splits, one opposite the other. They begin at opposite ends of the bush and finish a short distance from the other end. Larger bushes have more than two splits, all of a similar kind.

Bush **C**, however, is (in the writer's opinion) the best of the lot. It has a hole of clearance size drilled half-way along its length for the holder screw, which grips positively on the tool itself—the latter having a small flat ground on it for the purpose. The flange on the bush assists the tool-setter in removing the bush from the holder.

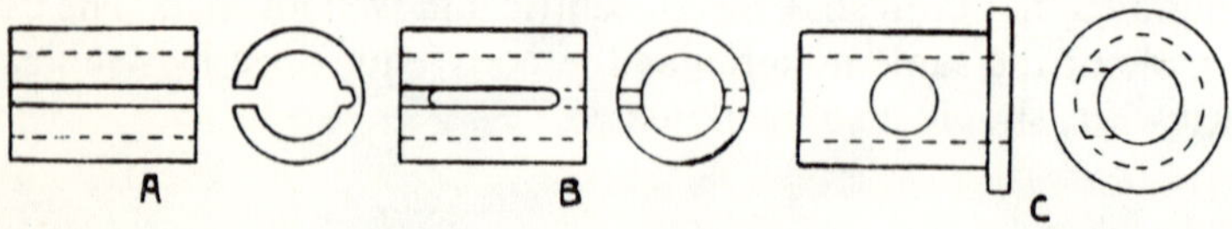

Fig. 6.15 Tool-holder Bushes

Whenever a holder screw binds direct on to any bush, the latter should always be provided with a flat lest the burrs formed by continual tightening of the screw make it impossible to remove the bush when that becomes necessary.

6.14 Floating Holders. The purpose of this type of holder is to help the tool avoid an error in working which can be caused by poor alignment of the machine spindle with the turret tool-holder or with the turret itself. The danger with some types of "floating" holders is that the "float" will not be parallel to the centre line of the machine. The top half of Fig. 6.16 illustrates what can happen.

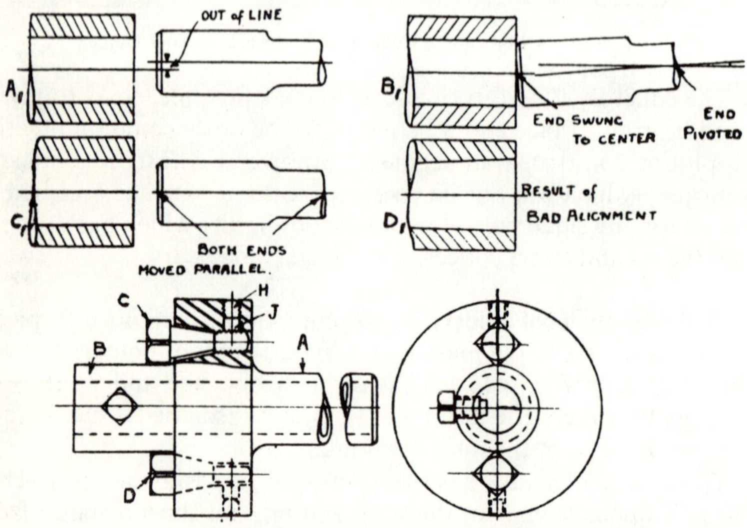

Fig. 6.16 Correct and Incorrect "Floating" (above): (below) an Efficient Floating Tool Holder

A_1 shows a workpiece in which a hole is to be reamed, and a reamer which is clearly well out of line. **B_1** shows a method of "floating" the reamer to bring it into line which is equally clearly wrong, because only the end of the tool has been moved. The result of reaming with the tool out of line in this way is shown at **D_1**, where the hole has been cut tapered, or (as it is commonly called) "bell-mouthed".

In **C_1**, on the other hand, the correct "floating" movement is illustrated, with the full length of the reamer brought parallel to the centre line of the machine. A simple pattern holder which achieves this is shown in the lower part of Fig. 6.16. It consists of two main parts, a shank **A** and a nose **B**, with the latter held against the former by the two screws **C** and **D**, in each of which a large part of the length is coned. It will be seen that the nose can either be locked firmly against **A** or be given varying degrees of "float" according to how far the coned screws are released. This will be done, of course, until the "float" is seen to be quite parallel.

The screws are locked by means of the small screw **H** bedding on the brass pad **J** to prevent them from working loose.

6.15 Tap-Holders. A good but simple design of releasing tap-holder has yet (in the writer's opinion) to be designed. Most of those on the market are both expensive and elaborate, and it seems unlikely that the same results could not be obtained with much simpler equipment.

Of the two types of tap-holder shown in Fig. 6.17, the upper is an

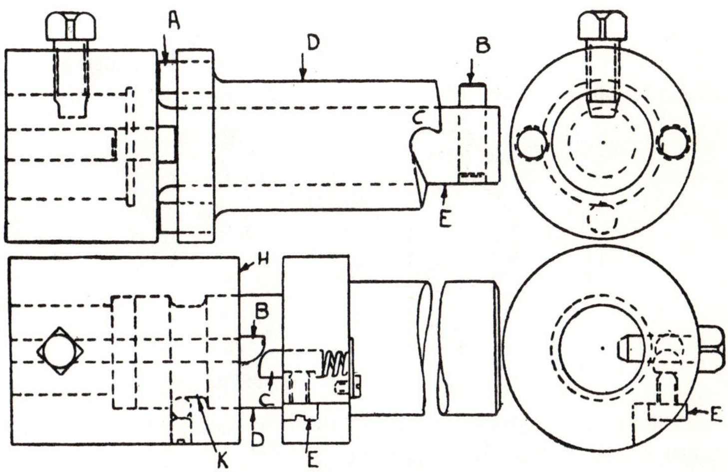

Fig. 6.17 Two Types of Tap Holder

old pattern whose chief faults are that wear is excessive, especially on the pins **A** and **B** and on the end of the catch **C**. Objectionable jarring also takes place when the pins disengage and when **C** engages, especially if the threads are fine. Another fault is that, when a tap of large diameter is being used, it becomes necessary for the sleeve **D** to be gripped very tightly. When this is done, it often happens that the sleeve closes slightly and grips the stem **E**—with results which will clearly be disastrous.

The best features of the holder illustrated in the lower half of Fig. 6.17 are the absence of vibration, and good accessibility to all its wearing parts. In the Fig., the nose is shown extended to its fullest extent. When its direction of rotation is reversed, the two pegs **B** and **C** engage quite smoothly, **C** being provided with a spring to take the thrust. To retain nose **H** on the shank **D**, a steel ball is inserted into race **K**. Peg **B** is fixed in its correct position, and peg **C** is given a flat which enables screw **E** to prevent it from revolving without impeding its sliding motion. In both pegs the tops are tooth-shaped, in order to assist release and enhance smooth working. Peg **C** is purposely made as simply as possible from standard drill rod, so that if it wears or breaks the tool-setter can replace it in a few minutes—a feature lacking in many other designs of tap-holder.

Note also that the hole into which peg **B** is driven should be continued absolutely straight through the nose so that the peg, once driven in, can be removed without too much difficulty when it becomes necessary.

6.16 Die Adapters. Die-holders working on the same principles as tap-holders can be readily bought or made, but it is also common practice to employ adapters of the pattern shown in Fig. 6.18.

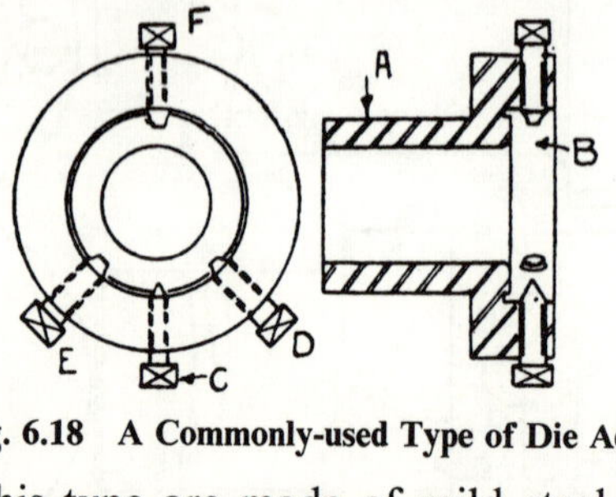

Fig. 6.18 A Commonly-used Type of Die Adapter

Adapters of this type are made of mild steel, case hardened to prevent damage. The shank **A** which fits into the bore of a standard tap-holder and the recess **B** which takes the button die (see Chapter 7) must both be made perfectly concentric, with the shank a good push fit into the tap-holder. The recess is improved by the narrow channel which can be seen to have been removed from its corner. Its purpose is to prevent dirt or cuttings lodging in the recess from causing the die to set at an angle.

Of the screws shown fitted in the end view, **C** has a coned point of some 60° whose purpose is to open the die, while the others, (**D**, **E** and **F**) help both to close the die and to hold it centrally. The die itself normally has small hollows ground into its rim for the screws to engage in.

Wear on these screws is heavy, so they should all be made of tool steel and heat-treated.

CHAPTER 7

Special-purpose Equipment

7.1 For a large number of operations standard equipment cannot satisfy all requirements; and to meet these requirements a wide variety of special-purpose equipment has been developed, of which a few examples will be described in this Chapter.

7.2 Truck Jigs. In Fig. 7.1, a multiple-drilling operation is shown being performed on a cylinder block—the latter not being visible because it is located inside the body of the movable, or "truck" jig. This method of machining heavy components enables a high rate of production to be maintained, for the truck is wheeled along to a variety of different machines where the necessary operations are performed on the component inside it. The disadvantage of the method is that the lines on which the trucks run and the absence of machine tables make it difficult to use on any other type of job the drilling or boring tools mounted to serve the truck jig, in the event of the workload on the truck slackening enough to make this desirable.

All fixtures of the truck type need to be very strong. The rails themselves are fixed directly on to the bed of the machine, so no machining difficulties arise because the truck is not running truly square under the spindles; but a small lateral float is allowed on the truck to enable the drills to find and enter the drill bushes easily. An indexing device is sometimes fixed to the machine bed plate on which the rails are mounted.

The bushed plates of the truck jig are located on two shouldered studs, and are then clamped by the knurled nuts. The body **A** is an iron casting, and has two trunnions which rotate in the end brackets **E**. The plunger for indexing is seen at **F**.

The rails on which the truck runs can be either of standard section or of solid rod (as in Fig. 7.1); and special supports or "chairs" are designed to hold them. The body of the truck is revolved directly by means of the handle **H**, but when the component is extra heavy a hand wheel with gearing will be needed instead.

Correct balance is of the first importance in the arrangement of the body casting, and the component itself as well as the jig plate needs to be considered in the calculation. An incorrectly balanced jig is a danger to the operator; and the designer should always specify on his drawing that the final balance is to be achieved by a trial-and-error process of adding or removing extra pieces of metal when the jig is completed, for it is impossible to guarantee a true balance by calculation alone.

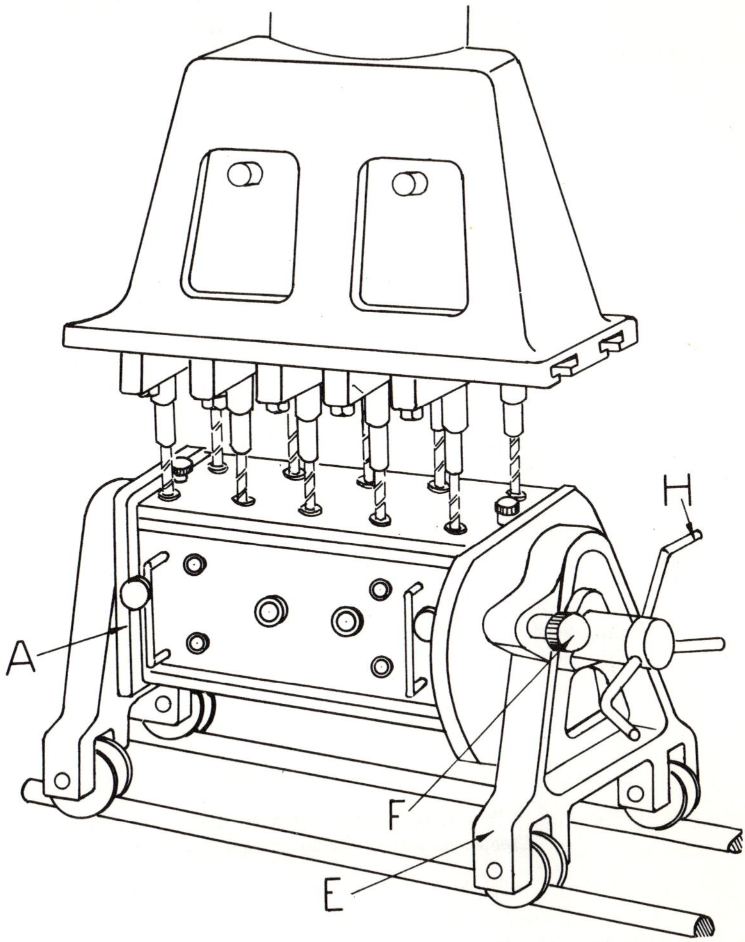

Fig. 7.1 A Truck Jig in position beneath a Multi-Spindle Drilling Machine

7.3 Grinding Fixtures. The purpose of the fixture illustrated in Fig. 7.2 is to grind the lip of the component seen in the foreground of the Fig., the three levers shown shaded being so ground at the same time.

In the first operation the angle piece is removed from base **X** and clamped to the table of the grinding machine in such a way that the surface **F** is ground at 90° to the main body of the lever. Note the two plugs provided for locating the levers from the reamed holes.

Subsequently, the 20° angle piece is replaced, and face **G** is ground at an angle of 20° to face **F**.

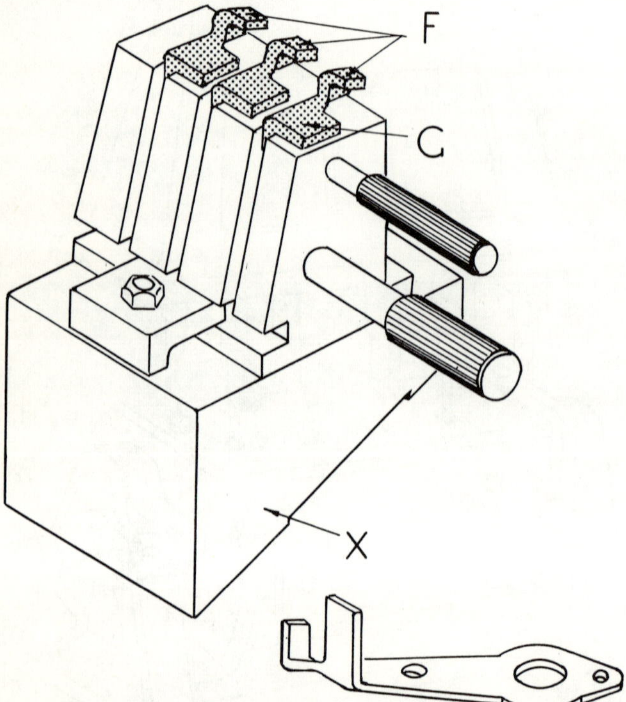

Fig. 7.2 A Fixture designed to Grind the Lip of the Component seen in the foreground

7.4 Coil-Winding Mechanisms. The diagrams in Fig. 7.3 illustrate a mechanism designed to wind the bobbins of electric bells. Basically, it is a form of lathe with provision made for gripping the bobbin and for controlling transport of the wire. **A** is a steel bar or bed on which slides the carriage **B**. Fitted to **B** are two threaded half-nuts **C** capable of meshing with either of the two lead screws **D**. **F** and **G** are hardened steel wedges which control the meshing of **C** and **D**. **E** is a hardened steel block of rhomboid shape fixed to the carriage **B**. The cradle **H**

carries two "V" guide wheels **J** and **K**, the latter being thick enough to take only the wire.

In operation, a spindle holding the bobbin drives two gears fitted to the lead screws so that they run at equal speeds but in opposite directions. As **B** is drawn along, **F** slides along one side of **G**. The rhombus **E** (see view at **X** on the right-hand half of the diagram) is inclined to the machine axis and so compresses the plunger **L**. Thus when **F** reaches the end of **G**, bracket **B** is forced to swing about **A**, disengaging one lead screw and engaging the other.

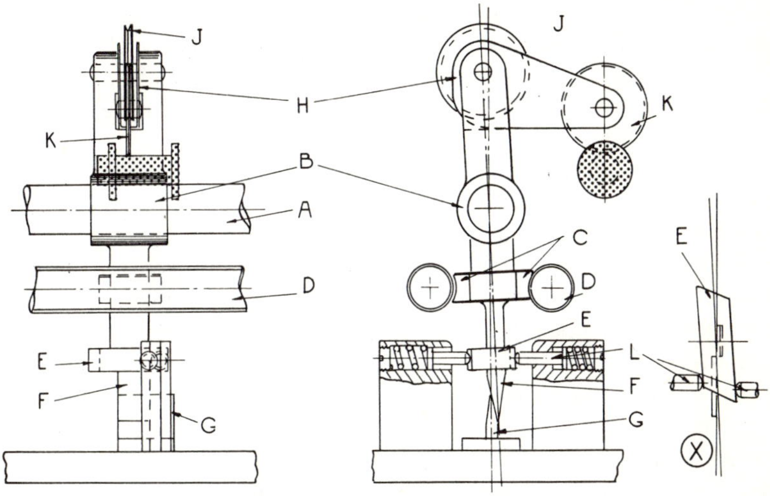

Fig. 7.3 A Mechanism for Winding the Bobbins of Electric Bells

This process can go on indefinitely; but if an adjustable contact is fixed on the top of cradle **H** to engage a micro switch, the drive can be stopped as soon as a pre-determined number of layers of wire have been wound on to the bobbin.

The distance of travel is governed by the length of the wedge **G** and by the positions of the side plungers **L**.

7.5 Magnetic Chucks. One of the tool designer's greatest difficulties is to achieve the efficient clamping or fixing of a component into its jig or on to the machine table. Often little or no provision is made on the component itself for the location of a clamp plate, with the result that awkward and expensive operations are necessary. Alternatively, the piece is often of so thin a section that it is impossible to avoid damaging it by normal clamping methods. An overriding necessity, of course, is that the method of clamping chosen must be cheap—economic, that is to say, in relation to the saving of handling time.

One of the quickest means yet devised for the loading, unloading and clamping of components is the magnetic chuck. Not only does it save a great deal of time; it also obviates the need for all spanners, levers and handles and so renders operation of the equipment as nearly fool-proof a process as possible.

Large castings, however, cannot usually be handled economically on the magnetic chuck, the latter being somewhat limited in size and power. With small components, however, its usefulness is great, particularly when the area to be machined lies close to the chuck face. The machining of areas lying more than a certain height above the chuck face presents difficulties because the greater leverage tends to turn the component over.

Magnetic chucks have been designed capable of exerting a downward pull of two mega-Newtons (2 MN) to the square metre, but even this pull is not enough to prevent some degree of lateral movement depending on the form of cutting tool employed and/or the method used to present the tool to the work. It is fatal if the movement of the machine table bearing the component against the cutter is too rapid, or if the thrust of a planing machine tool is too suddenly applied. If such faults can be avoided, the holding force provided by a strong magnetic field will do what no other method of clamping can do—and that is to exert a steady and powerful pressure equally over the whole base or foundation of the component.

Workpieces made of iron or steel are, needless to say, specially suitable for use with magnetic chucks, but articles made from non-magnetic materials can also be handled provided that iron or steel clamp-plates can also be introduced. Since this represents little improvement on the use of ordinary clamps in the first place, however, it is not a method of which common use is made.

When magnetic chucks are used, it must be borne in mind that the greater the mass of iron in contact with the chuck face, the stronger will be the magnetic pull. A magnetic field is almost always strengthened by the presence of iron, in that a larger number of lines of magnetic force will pass through a unit area of iron than would pass through the same unit area of air. Moreover, the softer the iron, the more permeable it generally is to magnetic lines of force.

A simple experiment will illustrate this fact. Place two magnets, each having one of its faces concave, in such a way that, when their opposite poles face one another, a cylindrical hole is formed (see

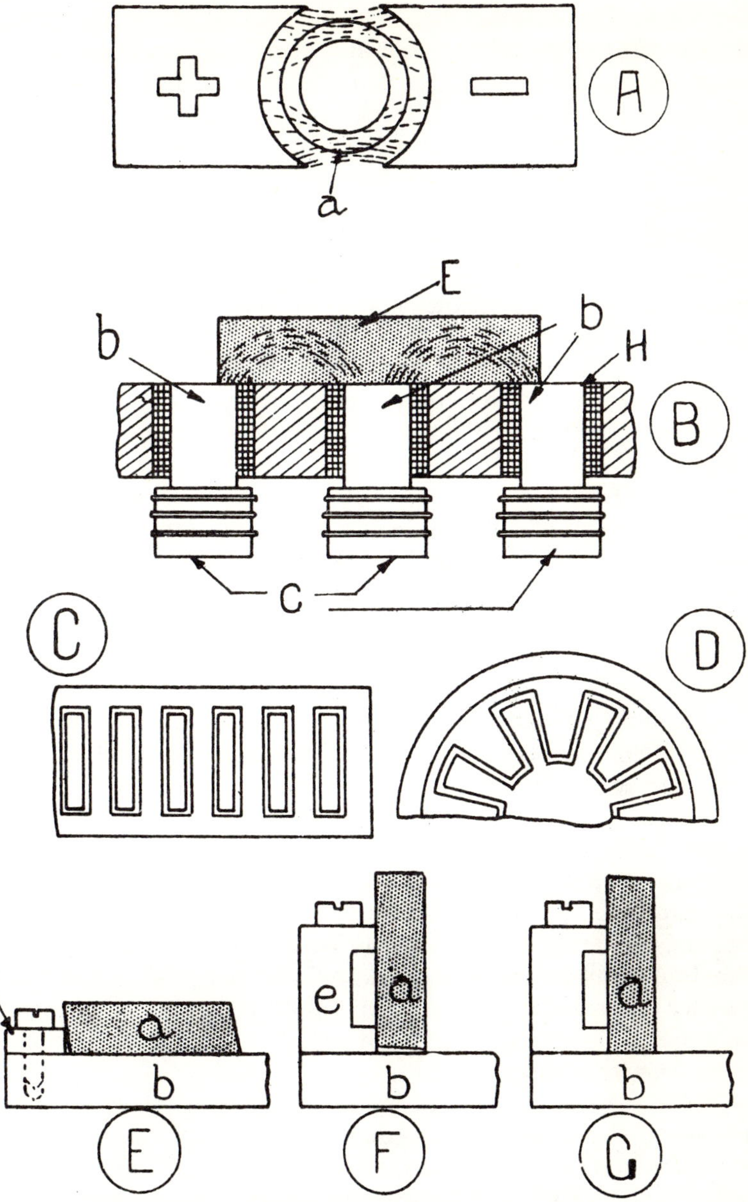

Fig. 7.4 *A*. **Lines of Force passing through an Iron Tube.** *B*. **The Principle of the Magnetic Chuck.** *C–D*. **Two Designs of Magnetic Chuck.** *E–G*. **Locating Work which is slightly out of square**

Fig. 7.4 **A**). The lines of force, if plotted, would be found to stream straight across from pole to pole. But when the iron tube *a* is introduced into the hole, the lines of force become distorted so as to crowd together through the tube in the manner shown, thus showing their preference for passing through iron rather than through air.

The basic principles of the magnetic chuck are shown at Fig. 7.4 **B**, where the face of the chuck on which the work is to be located is represented at **H**. Arranged about this face are a number of pole pieces *b*, which are magnetically insulated from one another by liners of white metal. Magnetism can be produced in these poles at will by allowing an electric current to flow round the coils **C**, which are wound round the poles. So when the work **E** is laid on face **H**, the lines of force pass from pole to pole through it and hold it firmly to the face. The greater the mass of iron, the more powerful will be the pull.

It follows from the above that the best gripping results are obtained when the work is of such a shape that as much of it as possible is in direct contact with the surface of the pole. An ideal arrangement would be if the poles could be arranged in any desired formation. While this is not wholly possible, there are on the market chucks having arrangements which can be adapted to many different needs.

The design shown at Fig. 7.4 **C** is one often used for the rectangular chucks employed on milling machines, and is well suited for the class of work generally performed on these machines. The circular chuck shown at Fig. 7.4 **D** is generally employed for small repetition work on vertical grinding machines, with the chuck rotating slowly beneath the grinding wheel.

The most efficient form of chuck (*i.e.*, the one which offers the maximum area of pole surface) consists of two main parts, a circular base and a covering cap. The cap has radial arms cast on it from which there branch off concentric projections. The base portion also is cast in such a manner that projections are left which fit the spaces between the projections in the cap, save only for a thin section of white metal inserted between base and cap.

Even in the machining of delicate pieces in which its usefulness is greatest, the magnetic chuck encounters special problems. Most mechanics know that all metals are in varying degree held in tension by their outer "skin", and that when this skin is removed, a movement of molecules takes place in the material itself which can have distorting effects. This distortion, known as "creep", is especially liable to occur when thin plate work is being machined.

Second, a constant flow of coolant must be maintained over a thin-plated or otherwise delicate workpiece lest heat generated by the heat of the wheel should cause it to warp.

Another trouble, more costly to correct, arises when a component arrives slightly warped before it is even placed on the chuck. When the current is switched on, the plate is gripped and made to lie perfectly flat, the magnetism overcoming the torsion. It is ground or otherwise machined in that state; but when it is removed from the chuck and the influence of the magnetism, the piece resumes its warped form. While this defect cannot in fairness be blamed on the magnetic chuck itself, it is nevertheless true that the chuck does tend to conceal the defect. The cheapest solution is always to have the components checked in the first place and any warped ones corrected or discarded.

For the grinding of short lengths of material of almost rectangular section in which one or more of the sides is slightly out of square, the method shown at **E**, **F** and **G** in Fig. 7.4 works well. At **E** the piece is shown lying on the chuck face *b* while its exposed face is ground flat, the stop strip **C** acting both as a means of location and as a stop to prevent skidding. One edge is then dealt with as shown at **F**, where the stop strip is replaced by the stop *e*. If the side of the strip already ground is located against this, the strip will be held perfectly vertical and the edge can be finished.

By using the same stop and arranging the work in a similar manner, the other edge can be finished as at **G**; while the grinding of the last face is accomplished with the aid of the same stop as was used at **E**.

When the edges of components of awkward shape require finishing, it is usual to give them support by sandwiching them between parallel strips, as shown at **A** in Fig. 7.5.

An effective method of handling rings which require grinding on their faces is indicated at **B** in the same Fig. The dial plate *a* is fitted in such a way that it can revolve closely over, but not touch, the small rotary magnetic chuck *b*. Below the dial is another disk which remains stationary, with its upper face on a level with that of the magnetic chuck. In the rotary dial plate holes are cut of such a diameter as to accommodate the rings to be ground, allowing about 3 mm of play in each. In the bottom plate, an escape hole is cut at *c*.

The method of operation is then as follows: The operator inserts a ring in the hole *d* and, with the current switched off, turns the disk round in the direction of the arrow so that a finished ring falls through the escape at *c* while an unmachined ring is presented to

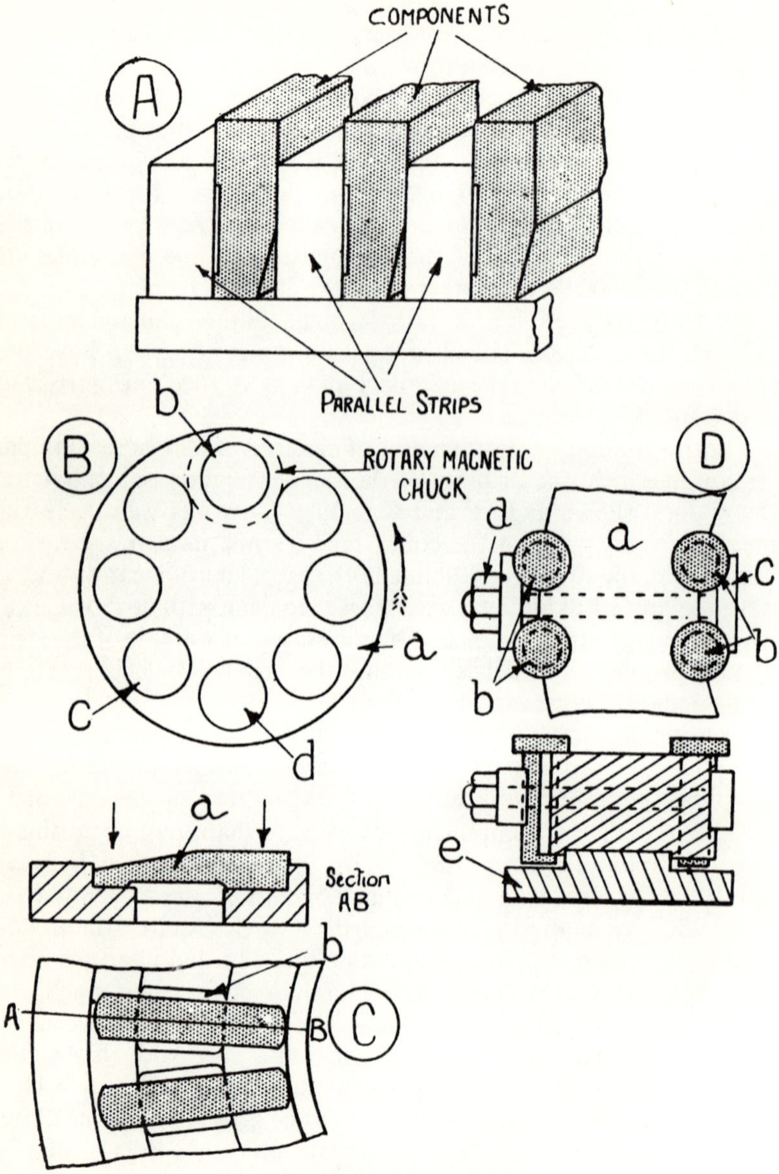

Fig. 7.5 *A*. **Locating Pieces of Awkward Shape.** *B*. **Grinding the Faces of Ring Components.** *C*. **Utilising the "Plasticity" of Magnetic Lines of Force.** *D*. **A Ring-shaped Fixture for Grinding Rod Ends**

the grinding wheel at *b*. A receptacle is arranged at *c* to catch finished components as they drop.

Shown at **C** in Fig. 7.5 is an instance of how the "plasticity", if such a term may be used, of magnetic lines of force can be made to serve a useful purpose. The components *a*, which require to be ground on their faces, are held in a number of radial grooves cut in a circular fixture, the latter being held on to the chuck by its own magnetism. To ensure that the components are held down quite flat, with no tendency to tilt by reason of an unequal magnetic pull, the work-holding fixture has a series of holes cut below the centre of each component, as shown at *b*. The permeability of the iron causes the lines of magnetic force to be diverted to the edge of the component, and it is successfully held.

It is in such ways as this that jigs and fixtures can be designed to hold the most awkward of pieces so that they can be machined accurately and without risk of being swept off the machine.

Another use for the magnetic chuck is found in the grinding of the ends of slender pieces such as valve tappets and rods generally. This operation is commonly performed with the aid of a circular fixture round which are drilled a number of holes close together to relieve the workpieces—with the magnetism of the chuck firmly gripping them all to the fixture.

If both ends of a stem need grinding, a ring-shaped fixture with a section like that shown at **D** in Fig. 7.5 will be found to work. The components *b* are located in the ring casting *a*, being held there by the draw bolt *c* which is locked by the nut *d*. With the piece held in this manner, it is possible to switch off the current and remove the fixture from the machine without risk of the components dropping off.

A fixture of this type is normally used in pairs, with one of the pair in operation while the other is being loaded close at hand. If a fixture for use on a particular job looks like proving exceptionally bulky, it can be made to consist of a number of pieces or segments if suitable provision is made for correct location of the pieces or segments, both on the magnetic chuck and in relation to one another.

Pieces of annular section such as bevel gears, etc., which require grinding on the face yet lack a base of sufficient area to provide stability, can be located on pegs arranged vertically in a circular plate. If these pegs are made of brass or bronze, the lines of force will be diverted through the iron of the components themselves to effect clamping.

7.6 Pneumatic Devices. Compressed air, usually at a pressure of 550 kN/m^2, is available in most modern workshops. It can be used to operate clamps and handling equipment, for the removal of swarf, and for the ejection and transport of small components after machining. A wide range of standard pneumatically-operated devices and control valves is available, and their cost is often justified in quantity production by the reduction in cycle time.

Note, however, that any system of pneumatic clamping must always be combined with some form of mechanical clamping, lest a sudden failure of the air supply should bring about the immediate release of the workpiece being held. This is a particularly important precaution when the components being handled are large or heavy.

The principal difference between an ordinary fixture and one intended to be operated by compressed air is that in the latter case the usual clamps are replaced by toggles or wedges operated by an air cylinder. Fig. 7.6 illustrates a typical construction.

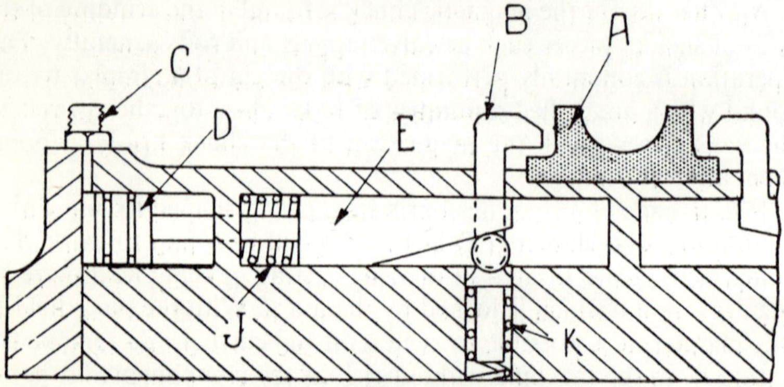

Fig. 7.6 A Typical Arrangement for Clamping by means of Compressed Air

The fixture is designed to hold the component **A** for a milling operation. The hook clamps **B** are made to grip or release the work. Air admitted through the cocks **C** forces forward the plungers **D**. To the other end of the plungers are fixed the wedges **E**, so that when the pistons are forced forward by the air pressure the wedges also are advanced. As this happens, the hooks **B** are drawn down and grip the work. The springs shown at **J** and **K** assist the hook clamps and the wedges to regain their open position when the air pressure is cut off. The admission of air into both pistons is controlled by a single valve, so that one movement of the lever either opens or closes both clamps.

The example given illustrates the elasticity of operation which the use of compressed air makes possible in a role of this type. If the clamping surface is uneven, clamps operated by a single piston, even if they all close at the same moment, will not all grip with equal force, whereas if each clamp is controlled by its own piston, full pressure will be exerted at every point every time the operating lever is moved to the "Clamp" position. Single-piston equipment, on the other hand, works well in operating draw-back or push-out collets (see Figs. 6.5 and 6.6) where the components to be held are of even profile and the jaws of the gripping mechanism can all bed equally.

Besides economy of operation—with any number of clamps capable of being controlled by the movement of a single lever—a compressed air installation has other advantages not possessed by a battery of magnetic chucks. The first is that materials of a non-magnetic nature cannot be gripped by the magnetic chuck, so that some form of special clamp is additionally necessary to handle them. Second, and of even greater importance, the removal of cuttings can be made automatic by arranging for a nozzle or nozzles to deliver compressed air at the right places, so saving the operator the work of washing or brushing them away. It is a disadvantage of magnetic chucks that dirt and cuttings tend to cling about both chucks and finished components under the influence of the magnetic fields, and their removal generally constitutes a separate operation.

CHAPTER 8

Cutting Tools

8.1 Materials. The material of which a metal-cutting tool is made must satisfy four requirements if a satisfactory tool is to be obtained:

1. It must have greater hardness than the material of which the workpiece is made if the cutting edge of the tool is to penetrate without distorting.

2. It must be sufficiently tough to prevent any break-up of the cutting edge under the impact of sudden shock.

3. It must be resistant to wear during cutting, so that tool life is reasonably long.

4. It must be capable of being ground to a sharp edge.

The principal materials in current use for the making of metal-cutting tools can be classified as follows:

(*a*) **High-carbon Steels.** The steel used in the manufacture of hand tools usually has a carbon content of between 0·75 per cent and 1·0 per cent; but in cutting tools such as drills, reamers and taps the carbon content is 1·0 per cent to 1·25 per cent. After the tool has been pre-formed, it is hardened and tempered to produce in it the required mechanical properties.

At high temperatures, however, carbon steel loses its hardness. It is therefore suitable only for tools which will be used for light work or slow-speed cutting. Its hardness at high temperatures, and also its tool life, can be improved by the addition to the steel of small percentages of alloying elements such as chromium, tungsten, cobalt and vanadium.

(*b*) **High-speed steels.** These are a range of air-hardening tool steels which allow much higher cutting speeds to be achieved. The most usual alloy currently in use contains from 0·5 per cent to 0·8 per cent carbon, 18 per cent tungsten, 4 per cent chromium and 1 per cent vanadium, and is known as 18-4-1 H.S.S. Molybdenum can also be used as the main alloying element.

High-speed steel is suitable for use in all types of cutting tool, its additional cost over plain carbon steel being justified whenever higher speeds or heavier cuts are possible. For rough machining, or for a cutting action which is intermittent as in milling, its toughness gives it an advantage over harder but more brittle materials. It is also fairly easy to form into intricate shapes such as drills, hobbing cutters and broaches.

(*c*) **Cemented carbides.** These are manufactured in small block form from sintered powders whose principal component is tungsten carbide. As their initial cost is high, cemented carbides are used either as tips for single-point tools which are then brazed on to a carbon-steel shank, or as inserts in multi-point tools such as milling cutters. They are extremely hard and resistant to abrasion, so they combine suitability for use at high cutting speeds with long tool life. They are therefore much used in turning tools and milling cutters on long production runs.

They are, however, rather brittle and tend to break up if the cutting action is interrupted. This tendency can be lessened by giving the tool a negatively-raked cutting face.

(*d*) **Ceramics.** The basic ceramic material used in cutting tools is sintered aluminium oxide. It retains its external hardness at high temperatures and has a low coefficient of friction, but it is very brittle and has low shock resistance. Its main application is therefore found in high-speed machining under light cutting conditions, *e.g.*, in the cutting of non-ferrous metals or in finish cutting. Ceramic tools are not suitable for use when the cutting action is intermittent; and they are easily damaged by the type of vibrations commonly known as "chatter", or when operating conditions generally are liable to sudden fluctuation.

(*e*) **Diamonds.** The diamond is the hardest material known to man, but its high cost means that its use must normally be restricted to light finishing work with single-point tools, and to grinding operations. For the latter, diamond grit is bonded with metal or resin to form the cutting face of the grinding wheel. A diamond tool—either single-point, multi-point or diamond-impregnated metal or resin—is also used for the "dressing" of grinding wheels to restore them to trueness after use.

The use of a diamond tool allows very high cutting speeds to be attained, and under suitable conditions tool life is long. The maximum benefit from the use of diamond tools is obtained when the machinery is sensitive, rigid and free from vibrations.

Resetting is usually carried out by the supplier rather than by the user of a diamond tool.

8.2 Cutting Fluids. The cutting action of a tool and the finish of a workpiece can be improved by applying a cutting fluid to the area where tool and workpiece come into contact. Such fluid is known as a "lubricant" or as a "coolant" depending on which of these functions is in any given case predominant. It is usually a soluble oil (emulsion), a mineral oil or a fatty oil.

Cutting fluids serve the following purposes:

1. To lower surface friction between the cutting tool and the chip by lubricating action. This reduces both the cutting forces and the power required, and also increases tool life.
2. To lower the temperature of both tool and workpiece by acting as a coolant. Tool life is thereby improved, and the reduction of distortion and thermal expansion results in greater dimensional accuracy. The workpiece is often rendered easier to handle by reason of its lower temperature.
3. To improve the surface finish of the work, principally by preventing or reducing the formation of a built-up edge on the tool.
4. To protect the work and the machine against corrosion.
5. To wash away chips and swarf from the cutting area.

8.3 Single Point Tools. The simplest form of tool has a single cutting edge; and this type of tool is used in turning, boring, shaping and planing operations.

The geometry of the cutting point (which largely governs the performance of the tool) is shown in Fig. 8.1, and the several terms and angles used therein must now be discussed.

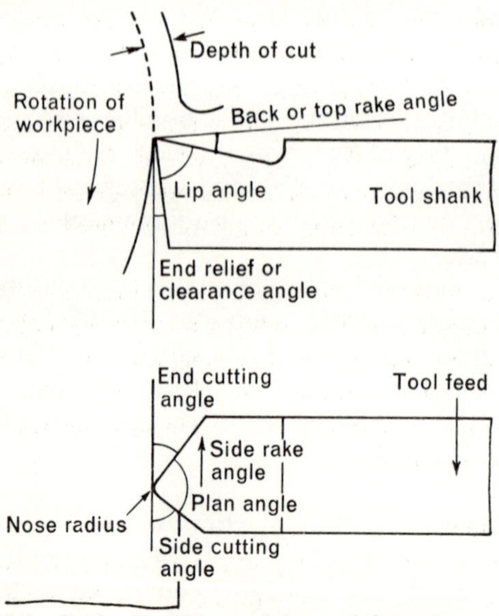

Fig. 8.1 Cutting Tool Nomenclature and Geometry

(*a*) **Rake angle.** True rake angle is always measured in the direction of chip flow across the face of the tool. It is therefore a combination of the *back rake angle* (which is measured in a plane perpendicular to the work axis) and of the *side rake angle* (which is measured in a plane parallel to the work axis). An increase in rake angle reduces the cutting force, so improving the surface finish but weakening the tool by reducing the *lip angle*. Low rake angles are therefore used in the machining of hard materials.

When the angle of the tool at the cutting point exceeds 90°, the rake angle is said to be *negative*. The use of negative rake angles with carbide or ceramic tools permits intermittent cutting action without great risk of damaging the tool.

Rake angle is usually between 5° and 20° for high-speed steel tools, and between zero and 10° for carbide tools. A negative rake of up to 10° can be used with either a carbide or a ceramic tool.

(*b*) **Relief angles.** To provide clearance between the tool and the cut surface, both end and side *relief angles* are supplied to the front and side cutting edges of the tool, respectively. They usually vary between 5° and 10°.

(*c*) **Nose radius.** To preserve the cutting tip of the tool, the nose is ground to a radius of approximately 3 mm.

(*d*) **Tool shank.** This should be of sufficient proportions to withstand the cutting forces, and stiff enough to resist deflections and the build-up of chatter. This is usually achieved by keeping *tool overhang* (the distance between tool-holder and cutting edge) to a minimum, and by setting the tool point at the same level as the axis of the workpiece.

8.4 Form Tools. The life of a cutting tool can be prolonged if it is allowed to retain the same contour at the cutting edge after every regrind. This is especially important in the cutters used in lathes and automatics, which are normally of three types—circular-form tool, dovetail-form tool or flat-form tool.

(*a*) **Circular-form Tools.** Fig. 8.2. shows an example of a circular-form tool for use in an automatic machine. Note that the form is spirally relieved at a lead of approximately 5 mm, so providing a positive cutting clearance on those parts of the tool which would be employed in turning the faces. Without such helical relief these positions would be rubbing, not cutting.

The cutting face is shown below the centre line in the illustration, the tool being set in the machine so that the cutting edge is brought up to the centre line of the work so as to provide front clearance. When this is done, however, the effective form of the tool is changed; and where great accuracy is required, a compensated form of tool is required.

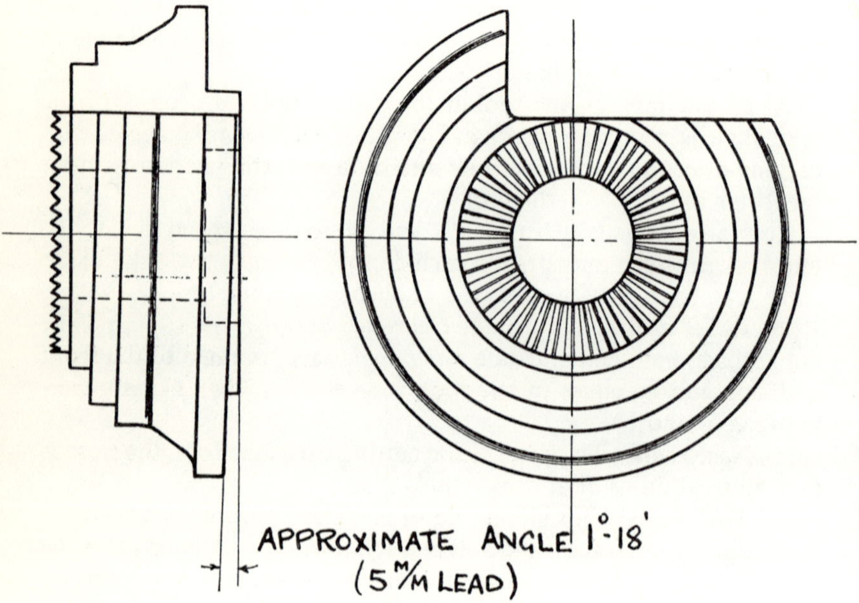

Fig. 8.2 A Circular-form Tool for use in an Automatic Lathe

The serrations visible on one side of the tool are standard, and serve to lock the tool securely into the tool-holder.

(*b*) **Dovetail-form Tool.** This type takes its name from the shape of the section running away from the cutting edge which enables the tool to be clamped in the special holder shown in Fig. 6.13.

(*c*) **Flat-form Tools.** The example shown in Fig. 8.3 is for use in a lathe. It consists of a high-speed-steel cutter mounted on a toughened steel shank by means of two screws. The cutter is held rigidly in position by a key fitted between the shank and the cutter. The tool is relieved to give a constant cutting form after every regrinding. In order to provide a top rake on the tool, a lip is provided which follows the contour of the cutting edge and leaves a "land", or flat surface, of about 0·12 mm on this edge. The tool is re-sharpened by grinding the top face on a surface grinder, and then regrinding the lip.

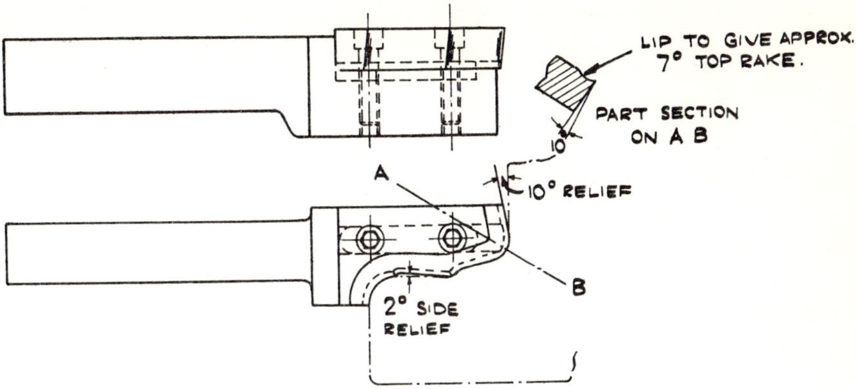

Fig. 8.3 A Flat-form Tool for use in a Lathe

8.5 Boring Tools and Holders. Although boring bars and tools have been the subject of a number of recent patents, many of the older types of bar are still in constant use. One of the simplest is shown in Fig. 8.4 (*a*), where the tool **A** is inserted in the bar **B** and held by means of the set-screw **C**. To facilitate adjustment and to prevent the tool pushing back when under heavy cut, the screw **D** is set behind the tool.

The point of arranging the tool across the bar at an angle is to enable a slightly longer tool to be used.

If necessary, the finishing tool can be arranged close behind the rougher, in which case it should be so located that the cutting edges of the two tools are at opposite sides of the bar so as to equalise the cutting strain.

When the hole being bored is small and the bar is being kept as large in diameter as possible, the presence of a binding screw on the circumference may not be possible. This drawback can be overcome by locating the screw at the end of the bar, as shown in Fig. 8.4 (*b*).

Fig. 8.4 (*c*) shows another method of holding a boring tool, the length of the latter equalling the diameter of the hole to be bored. A flat is made on the side and in the centre of the tool, and a taper peg *a* is driven in. To give a rough idea of the proportions of this type of cutter to the boring bar, it will be found that a satisfactory value for the diameter of the cutter (or its thickness if it is not of round section) is about one-fifth the diameter of the boring bar.

Figs. 8.4 (*d*) and 8.4 (*e*) show two other methods of fixing cutters. In the former the bar **A** is accurately slotted to take the tool **B**, the latter being held in place by means of the tapered screw **C**. This taper comes in useful when it is desired to "float" the cutter, for it is then only necessary to release the screw sufficiently to give the amount of "float" desired.

In Fig. 8.4 (*e*), a portion is milled out of the centre of the cutter *a* which enables it to be located by means of the wedge *b*.

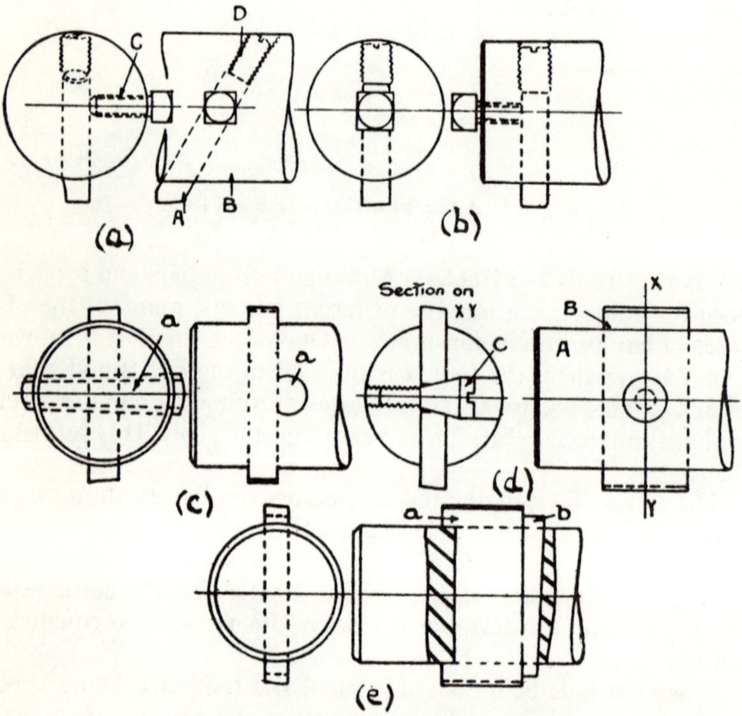

Fig. 8.4 Boring Tools and their Holders

Both the cutting and the clearance angles for boring tools depend on the nature of the material being cut, and in some cases also on the section or form of the component being bored. Fig. 8.5 shows at **X** and **Y** a roughing and a finishing tool, respectively. The former, it will be noticed, has a pronounced cutting edge *a*, given it by the tool angle of 60°. This provides a good free-cutting action, but the feed must not be too coarse. If properly heat-treated before use, such a tool will have a long life.

Clearance is provided on the side *a*, the top *c* and the end *b*. At *a* and *b*, this clearance should be about 4° for cutting cast-iron and brass, increasing to 6° for malleable iron and soft steel and to 10° for aluminium. The end clearance need only be some 2–3°, and some designers leave a finishing tool with no clearance at all on its end, and only a few degrees on its side. In the writer's opinion this is not advisable, for considerable friction will arise and particles of metal will sometimes stick on the cutter and cause enlarged bores.

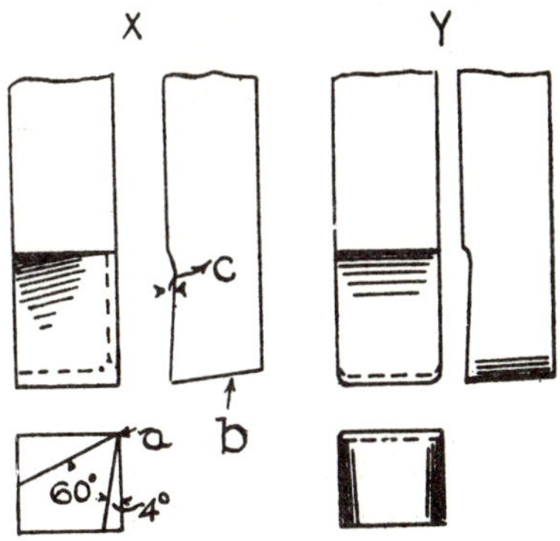

Fig. 8.5 Boring Cutters: a Roughing Tool (X) and a Finishing Tool (Y)

It is the accuracy of the fit of the boring bar in its guide bush which governs the final accuracy of the work produced, so no method of maintaining this accuracy of fit should be rejected. Fig. 8.6 illustrates one such method, in which several slots are milled along the length of the bar, equally spaced round its circumference. Into these slots are fitted hardened steel strips. The bar is then placed between centres in a grinding machine, and the strips are circular-ground to the required diameter.

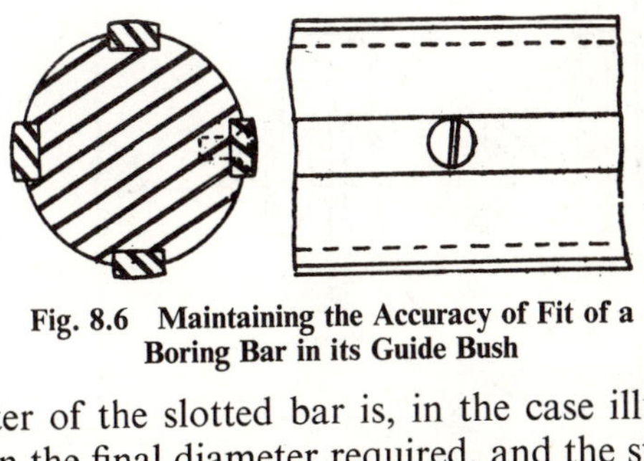

Fig. 8.6 Maintaining the Accuracy of Fit of a Boring Bar in its Guide Bush

The diameter of the slotted bar is, in the case illustrated, some 3 mm less than the final diameter required, and the strips stand out about 2·5 mm above this before they are ground.

8.6 Boring Cutters of Special Form. Some workpieces call for a cutter of special irregular form, or else one having several steps of different diameters whose function is to bore a hole in which these diameters are concentric and at given distances apart. Fig. 8.7 illustrates a

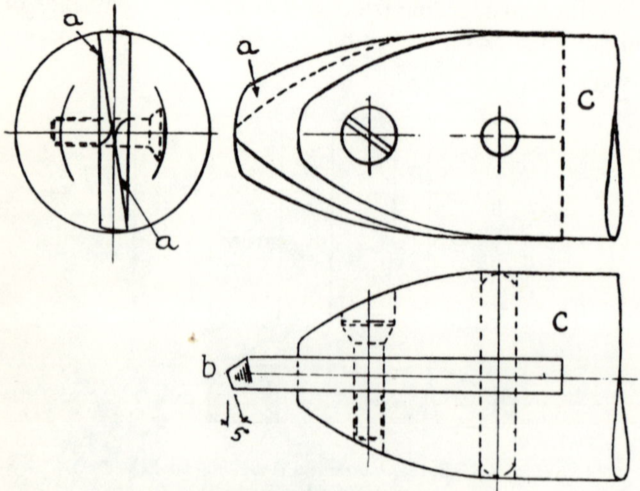

Fig. 8.7 A Flat-Bladed Boring Cutter used to shape the interior of projectiles

special flat-bladed boring cutter which has been used to form the interior of certain projectiles. Top rake is obtained by grinding the cutting edges as shown at *a*. This also allows the edges to lie central for the purpose of machining a flat bottom when necessary. The front clearance at *b* is about 5°.

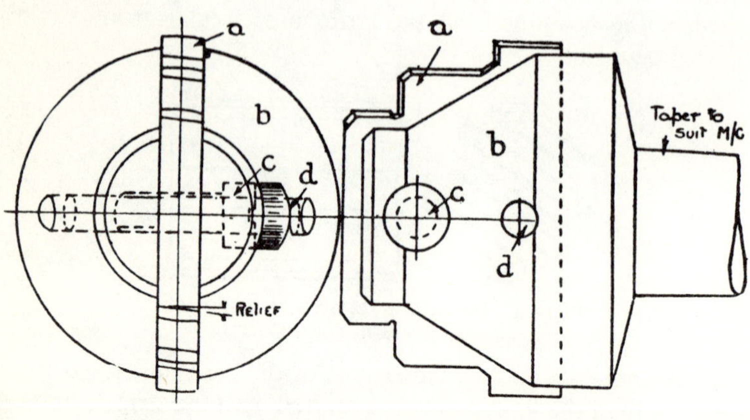

Fig. 8.8 A Boring Head holding a Three-Stepped Cutter

For roughing purposes, the profile of the cutter is given a number of nicks or chip breakers. The cutter itself is located centrally on the bar *c* by means of a slot milled out to form its seating. It is held there by a screw and a dowel pin as shown.

A boring head holding another special cutter is shown in Fig. 8.8. Here the cutter *a*, which has three steps, is held in a slot in the head *b* by means of the cheese-headed screw *c* and the dowel pin *d*.

Whenever possible, a component having a bore consisting of several steps should be machined on the capstan lathe, with separate tools being used for each diameter. The reason is that the manufacture of specially formed blades is expensive, and they are usually fit only for scrap when worn undersize. Though taken from actual practice, the boring head illustrated in Fig. 8.8 cannot be recommended as good tool engineering for that reason.

8.7 Flat Bits. Flat bits of the type illustrated in Fig. 8.9 can sometimes be used in place of the reamer and counterbore, their great advantage being cheapness of manufacture. During World War II, very large numbers of tools of this type were used on fuse work, for it was an easy job to turn the blanks on a good capstan lathe and to set up an indexing head on the milling machine to mill the two sides. There then only remained the three important operations of giving side clearance, hardening, and finally grinding to size.

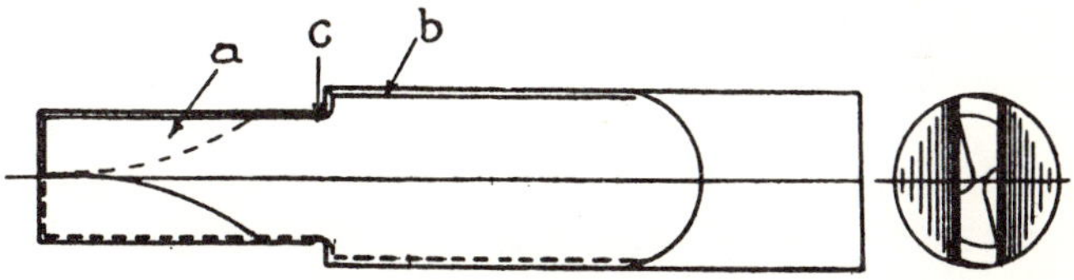

Fig. 8.9 A Flat Bit

In Fig. 8.9, the portion *a* performs the work of an ordinary reamer, while the rear part of *b* does the work of a counterbore. The blade is undercut as shown in order to enable the cutting edges at *c* to be so ground as to produce a square shoulder, and also so as to facilitate subsequent grinding. The blade portion should be of reasonably thick section—about one-fifth of the diameter of the cutter is generally satisfactory. When there are two diameters, as in Fig. 8.9, the relevant diameter can be taken as the mean of the two.

The front clearance for the counterbore, and for the bottom of the reamer if it has to make the bottom of the hole square, should not exceed 4° lest the cutter tends to dig into the work and risk breaking.

Bits and reamers have a general tendency to produce holes which are "bell-mouthed", or larger at the mouth than they are at the bottom. To counteract this, a slight back taper on the bit or reamer of some 0·4 mm per metre will be found an advantage.

8.8 Counterbores and Spot-facing Cutters. There is not much difference between these two types of cutter; but while a counterbore can be used for spot-facing, a spot-facing cutter cannot be used for making a counterbore. Inspection of Fig. 8.10 will quickly show that the cutter at **A**, which has no side teeth, could not contrive the escape of cuttings once it had penetrated far below the surface of the work.

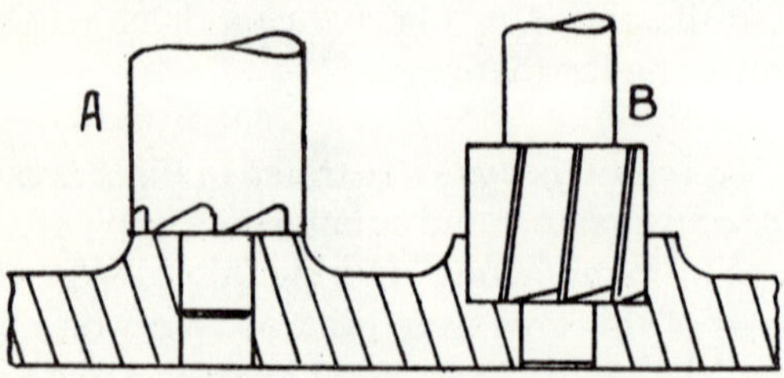

Fig. 8.10 A Spot-facing Cutter (A) and a Counterbore Cutter (B)

The teeth of both counterbores and spot-facing cutters can be either spiral or straight depending on the nature of the material to be machined. In counterbores over 15 mm in diameter, the shank can be manufactured separate from the cutter portion, with the latter being slipped on and retained by a pilot pin screwing into the shank.

The angle of spiral should be about 5°—an amount which will also be satisfactory for front clearance on the teeth. Since a common trouble with piloted cutters is the seizing of the pilot pin in the hole, this pin is given a small groove to retain any grease, and also to collect cuttings which could otherwise cause binding. It is for this reason that some form of groove or slot should be provided on all bars or pins which are used for steadying or guiding purposes.

8.9 Twist Drills. The basic design is illustrated in Fig. 8.11.

Though twist drills are nowadays both specialised and standardised in design, it is essential that the jig-and-tool man should thoroughly understand the reasons why they are given their exact form; for not only will he sometimes be called upon to specify on his tool layout the type of drill he requires to cut the particular kind of material being handled, but he may also have to design into a combination tool a part which will perform all the functions of an ordinary drill.

All cutting edges must be ground at the correct angle to the axis of the drill. For general purposes, an angle of 59° has been found most suitable. Both cutting edges, too, must be of equal length.

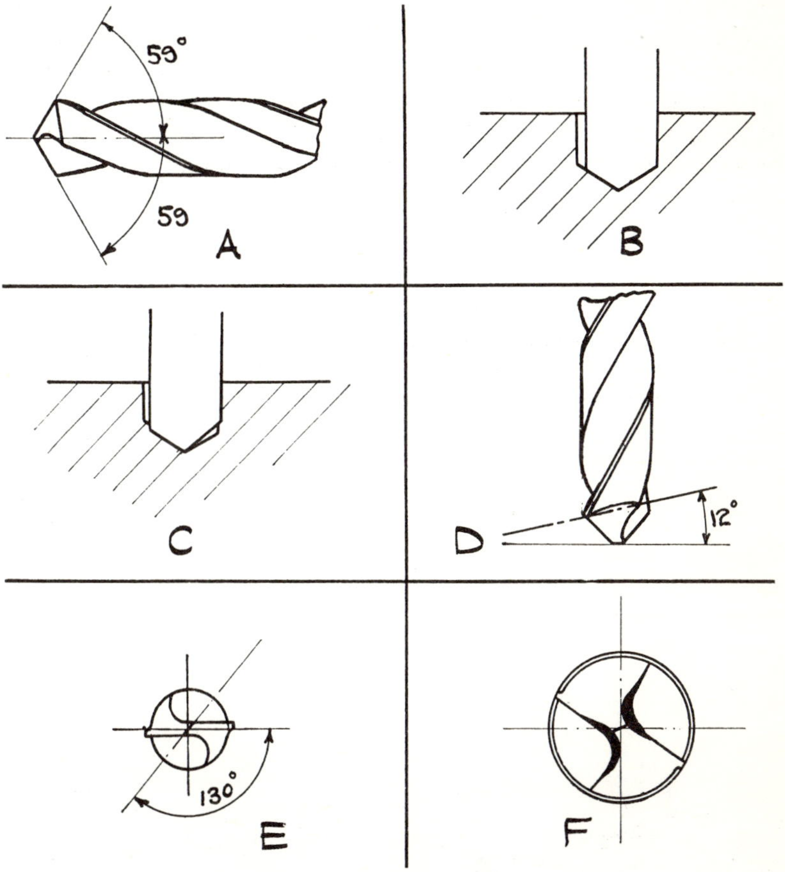

Fig. 8.11 Basic Twist Drill Design Features (*A*, *D*, *E* and *F*) Errors Due to Inaccurate Grinding (*B* and *C*)

In Fig. 8.11, **A** shows a drill correctly ground; while at both **B** and **C** it will be seen how impossible it is to obtain accurate results with inaccurately ground drills. At **B** the drill has cutting edges of equal angle but of different length; the point of the drill is consequently out of centre, and an oversize hole will inevitably result. At **C**, on the other hand, while the point is central the cutting edges are not of the same angle—with the result that all of the work must fall on only one lip.

Correct clearance angle on the periphery of a twist drill is 12°, as shown at **D** in the Fig. The centre angle which will be found to give the best results when combined with a 12° clearance is 130° (**E** in the Fig.).

The central core of a drill is known as the *web*, and one of the principal reasons why drills split up the web, particularly when they are heavily loaded, is insufficient lip clearance. When heavy feeds are to be taken on soft material, improved results can often be obtained by increasing by one or two degrees both lip-clearance angle and the centre angle.

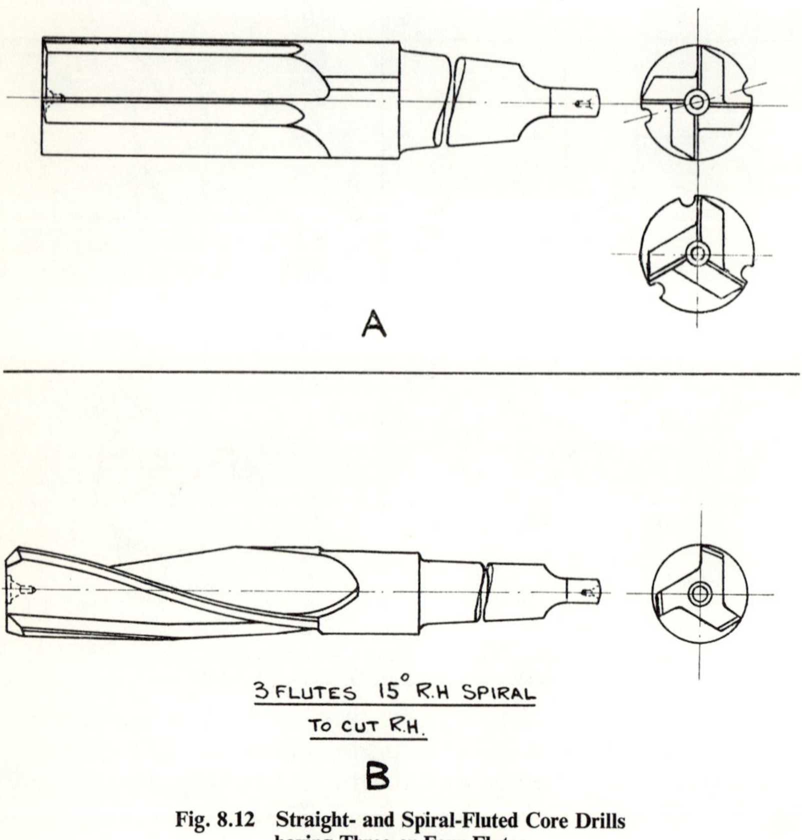

Fig. 8.12 Straight- and Spiral-Fluted Core Drills having Three or Four Flutes

8.10 Core Drills. Fig. 8.12 illustrates three examples of core drill heads such as might be used for the opening up of holes which have been either rough cast or roughly pierced, as in the case of stampings. When accuracy is required, core drills are made 0·4 mm undersize to allow for a subsequent reaming operation. They need to be piloted through drill bushes, and for this purpose their outside diameters are ground circular and a substantial "land" is left on that part of the outside diameter which remains after the flutes are milled. On a 50 mm drill, about 3 mm should be allowed for "land". Being front cutting, core drills need only to be reground on the front edges.

The two sections of core drills shown at **A** in Fig. 8.12 are suitable for cutting cast-iron pieces, while the spirally-fluted type illustrated at **B** is more suitable for steel. It has also been found to work better on elektron metal than does the straight-fluted type.

8.11 Core Drills for Finishing Bores. Should the work call for better finish and alignment, the hole may be drilled smaller than is required and a core drill similar to that illustrated in Fig. 8.13 may be fixed to the end of a body of suitable length and fed through. The tool shown has three cutting edges and is piloted in a true path by means of three brass pads let into the drill, screwed and finally soldered into position.

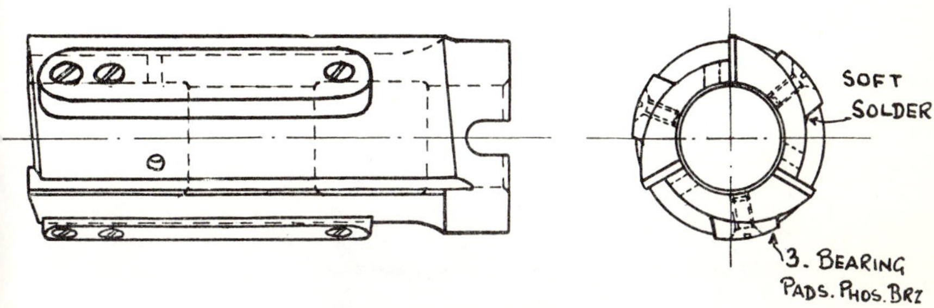

Fig. 8.13 A Core Drill used for Finishing a Long Bore

An alternative design for such a tool is shown in Fig. 8.14, in which three inserted blades are clamped in the drill body. One of the three pads is made from lignum vitae, and is left 0·25 mm larger than the diameter which the blades are to cut. When the lignum vitae enters the space newly created behind the cutting blades, it is scraped to size by the end of the hole, so holding the brass pads closely against the inner wall.

Lubrication is carried to the cutting edges through the centre of the drill, and by way of small holes drilled from the outside diameter into the centre hole. To ensure that this tool (and the one shown in Fig. 8.13) is given no element of bias when starting, the hole required should be bored to size with a single-point tool for a short distance along its length before the core drill is inserted.

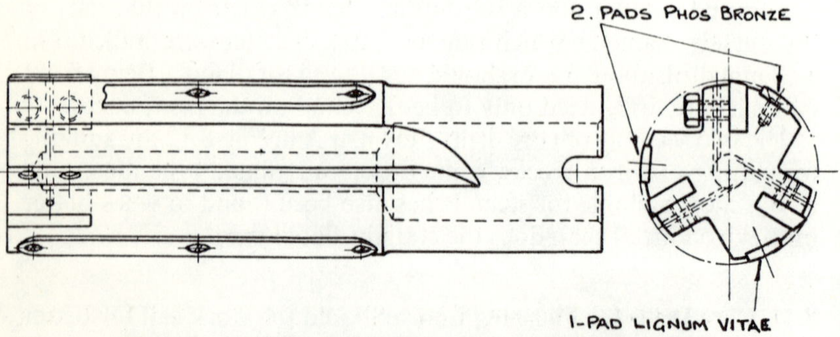

Fig. 8.14 An Alternative Design of Core Drill for Finishing Bores

8.12 Reamers. A reamer is essentially a finishing tool used to machine holes accurately to size after a drilling or rough-boring operation. Although its method of operation is similar to that of a drill, the reamer cuts only on its circumferential lands, removing not more than 0·75 mm of metal from the hole.

Reamers can be used either by hand or in a machine, and may be either parallel or spiral fluted. There are also expanding reamers, shell reamers, and other special-purpose tools.

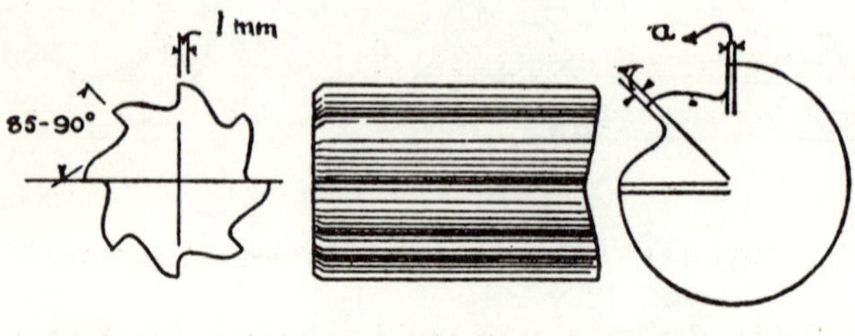

Fig. 8.15 Tooth Form in a Standard Parallel-Fluted Reamer

Fig. 8.15 shows the cutting teeth of a standard parallel-fluted reamer. Special milling cutters are used to form the flutes, the angle between neighbouring teeth being between 85° and 90°. Such an angle gives a substantial form of tooth, and at the same time provides a groove of sufficient size to facilitate the escape of cuttings.

Sometimes the reamer teeth are cut with their top faces slightly ahead of radial, by the amount *a* indicated in Fig. 8.15. This amount should be about 0·4 mm for reamers up to 20 mm diameter, and 0·7 mm for larger sizes. Sufficient land must be left to act as a steady for the reamer, the width being about 1 mm as indicated. This will serve to prevent chatter, and to improve the accuracy of hole size and shape.

It is also claimed that chatter can be prevented by uneven radial spacing of the teeth; but if this is attempted, identical teeth must be cut diametrically opposite one another.

Many reamers, standard and special alike, have teeth of either left-handed or right-handed spiral, it being claimed that the shearing cut produces a better finish. A left-hand spiral certainly prevents the reamer being drawn into the work when hand fed; but with machine reamers this point does not arise, and a right-hand spiral which assists the withdrawal of chips from the reamed hole will generally be found superior.

8.13 Expanding Reamers. Many ideas, most of them patented, have been advanced with the aim of saving the expense of scrapping reamers when they become worn undersize. A simple but successful example is shown in Fig. 8.16.

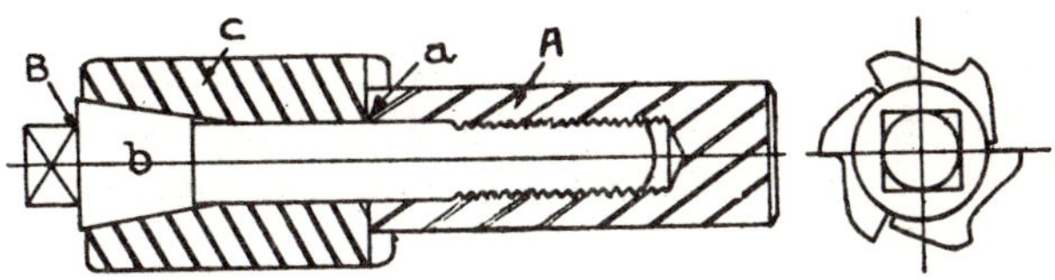

Fig. 8.16 A Simple Form of Expanding Reamer

It consists of a stem **A**, a screw **B** and a shell portion **C**. The stem, made of mild steel, has a tongue at *a* which engages in a slot milled in the bottom of the reamer **C**, so giving a positive drive; and a hole is bored and tapped to receive the screw **B**. The latter is made of case-hardened steel and has a thread of fine pitch to stop it working loose. The coned portion *b* is ground after hardening.

The shell portion **C** of the reamer is of high-speed steel bored to fit the plain part of the screw **B**, and it has a female cone of 15°. Three slots equally spaced round its circumference permit expansion, breaking into the bore and terminating some 10 mm from the rear end. When the screw **B** is turned with the aid of its square end, its male cone forces open the reamer.

The expansion thus provided is admittedly not parallel, since only the front of the reamer expands. But this does not matter because a back clearance or taper is created which is, in some classes of work, an actual advantage.

8.14 Special-Purpose Reamers. Though standard tools, which are cheaper, should be used whenever possible, a tool of special design is sometimes essential. Fig. 8.17 illustrates such a case, where a reamer and a counterbore are shown combined in a single tool—a device often employed on a turret lathe when more tools are needed than the turret can accommodate.

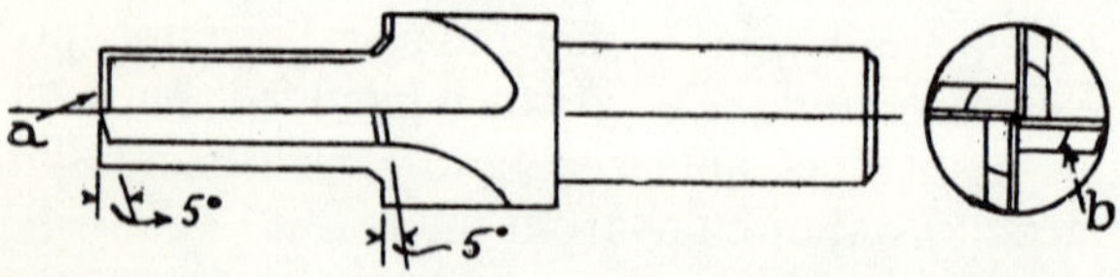

Fig. 8.17 A Reamer and Counterbore Combined in a Single Tool

A combination tool of this kind is best avoided, however, if the hole in question needs to be reamed to very accurate limits, for if the front cutting edges of the tool are unevenly ground, or if they are allowed to become dull, the cutter will tend to spring away from the work, so enlarging the hole.

A method of strengthening the teeth of the reamer, and of preventing chatter, is to relieve, or "back-off", the cutting blades as shown at *b* in Fig. 8.17, where the outer edge is made convex in form instead of being left flat.

The clearance on the front end of the special-purpose reamer described, and on the teeth of the counterbore, should not exceed 5° lest the tool "dig in" and possible break. The particular cutter shown can be used to ream a "blind" hole (*i.e.*, one blocked at one end) and also to make the bottom of a hole truly square—this latter job being performed by the teeth cut on the front face of the reamer.

8.15 Roller Milling Cutters. These are cutters of cylindrical form which are caused to rotate each about its own axis while the work moves under them on the machine table. They are used to machine flat plain surfaces. Fig. 8.18 illustrates one form of roller milling cutter, such as would be used for heavy work on broad surfaces.

Given the outside diameter of the cutter (it should be made as small as possible without making the shell too weak), the next point to consider is the number of teeth. For facing or "slabbing" work, where the cut is generally heavy and where the escape of cuttings from under the cutter is difficult, a coarse-pitch cutter with few teeth and plenty of space between each of them is to be preferred. It is true that the cutting loads would be reduced by dividing the work between a greater number of teeth, but it must also be remembered that the finer the pitch, the more teeth will be in action at one and the same time. The principal objection to fine-pitch cutters, however, is the lack of room for the escape of cuttings. If sufficient "clearance" or "chip capacity" of this kind is not provided, resistance to the cutting action, and consequent friction, will develop.

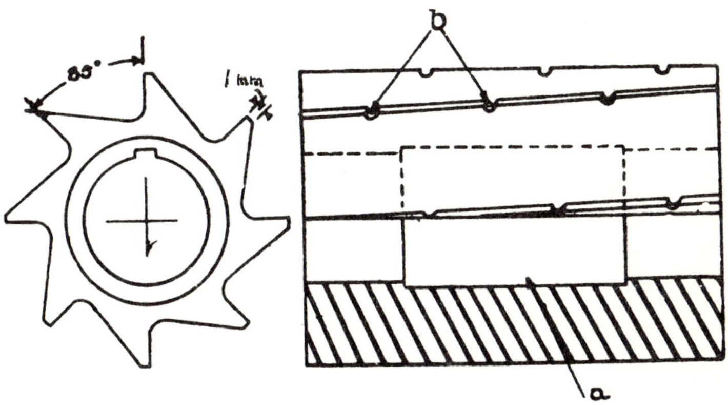

Fig. 8.18 A Roller Milling Cutter

At the same time, the attainment of adequate chip capacity must not be achieved at the cost of seriously impairing tooth strength. A tooth of adequate strength will be obtained by allowing about 0·4 mm for land at the cutting edge and making the clearance angle about 85°, as in Fig. 8.18. The land referred to, of course, has a back clearance of between 2° and 5°, in addition to the clearance angle of 85° mentioned.

With the increasing use of carbide-tipped tools, together with more robust machine-tool construction, it has been found that cutters having a negative rake can be made to give superior results. The power required is considerably increased but the added tooth strength is an advantage, especially on the heavier grades of work.

A spiral tooth is certainly preferable to a straight one since its cutting action is greatly improved by the more gradual engagement of each succeeding tooth. This is especially beneficial when the material to be cut is of a tough fibrous nature. The angle of the tooth helix varies considerably in tool design, but is normally between 10° and 15°.

In all milling cutters over 20 mm wide, the bore is undercut as shown at *a* in Fig. 8.18, the length of the undercut being between one-half and two-thirds the width of the cutter. The reason for this is manufacturing convenience. The undercut facilitates the grinding of the bore of the cutter, and does not impair its strength.

Cutters which are long in relation to their diameter can be given *chip breakers*, as shown at *b* in Fig. 8.18. These are concave nicks on the face of the cutter whose depth is about twice the width of the land and whose purpose is to relieve cutting loads by limiting the width of individual chips.

8.16 Inserted Tooth Cutters. The diameters of all milling cutters, of whatever type, should be kept as small as possible, because the life of these tools is limited and there is always the risk of the cutter breaking in the final hardening process. Since steel is expensive, the less there is of it at risk in either of these ways, the better.

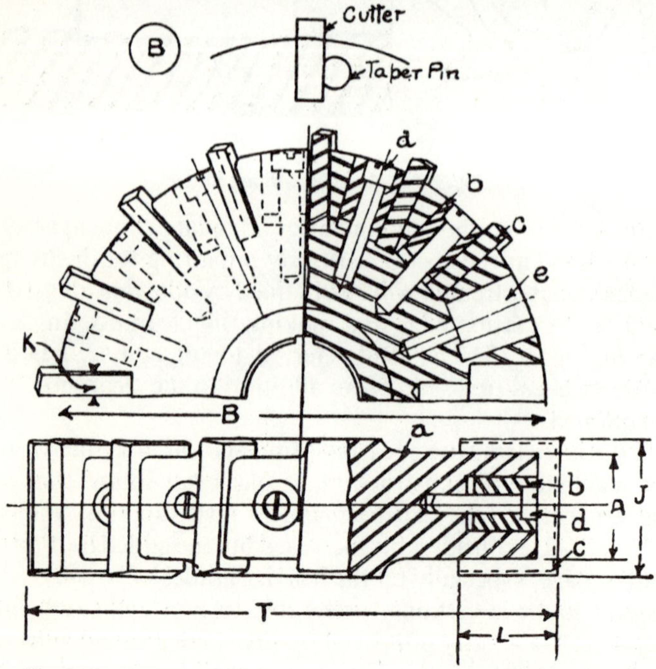

Fig. 8.19 A Large-Diameter Cutter of Inserted-Tooth Design

When a cutter of large diameter is essential, it is best to use one of the inserted tooth type illustrated in Fig. 8.19.

The body *a* of a cutter of this type is generally made of mild steel, and the bushes *b* of tool steel. These bushes have a taper of about 2° and so act as a wedge holding the cutter blades *c*. These are either of high-speed steel or of mild steel tipped with tungsten carbide. The screws *d* which hold the bush in its seating are also made of tool steel to give them the required strength.

The holes *e* into which the bushes fit have a taper corresponding to that of the bushes. The latter also have a flat designed to enable them to grip the cutter blade to maximum effect.

8.17 Relieved-Form Cutters. When cutters are relieved or backed off in the ordinary way, their form will be altered as the cutting face is ground away. But it is frequently essential that the form of a cutter be accurately maintained throughout its life, and form-relieving is a way of doing this while still giving the necessary clearance.

Form-relieving is accomplished on a special lathe (or similar machine), and it is incumbent on the designer to indicate on his drawing that this method is to be used. Fig. 8.20 illustrates a case in point, with the relief clearly indicated.

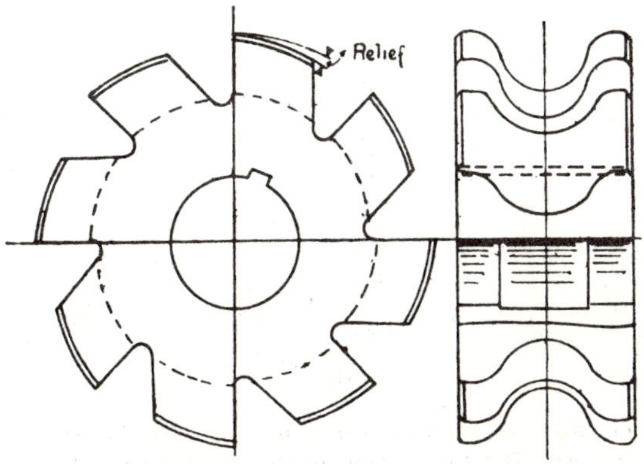

Fig. 8.20 Form-Relieving to Prolong Accurate Cutter Life

It is important that the relief be not excessive lest too weak a tooth result; yet if insufficient relief is allowed the cutter will rub rather than cut, and will quickly be destroyed. A reasonable relief is obtained by striking an arc from a centre offset one-eighth to three-sixteenths of cutter diameter from the true centre of the cutter.

Note that the clearance angle between the teeth in Fig. 8.20 is only some 45°. This allows an increase in the circumferential length of the tooth, and so prolongs its life. Since form cutters seldom need to take heavy cuts, clearance angle to give chip capacity need not be so great as it was in the roller mill discussed in Section 8.15.

Form cutters are often provided with chip breakers similar to those used on plain cutters.

8.18 Broaching. Broaching is a machining process used to fashion either internal or external surfaces of a uniform transverse section in a single pass of a long multi-point tool. This broach, as it is called, is drawn either through a hole or longitudinally over a surface.

Fig. 8.21 illustrates three different types of broach. Each consists essentially of a straight core having a series of protruding teeth which form the cutting edges. The teeth are usually made of high-speed steel, either in one piece or in sections; but for long production runs or in cases where close tolerances are required, inserted teeth of cemented carbide are common.

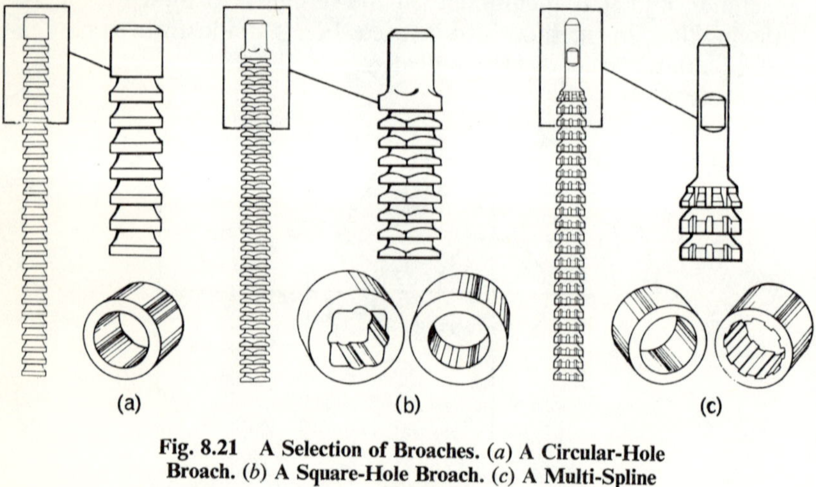

Fig. 8.21 A Selection of Broaches. (*a*) A Circular-Hole Broach. (*b*) A Square-Hole Broach. (*c*) A Multi-Spline Broach

Every section of the work is cut by all the teeth on the broach at every stroke. This is achieved by gradually increasing the circumference of either the teeth or the core, with the smallest teeth or core diameter placed at the starting end of the broach.

There are three categories of teeth: *roughing* teeth which remove most of the metal, *semi-finishing* teeth which complete the shaping, and *finishing* teeth. It is desirable, in order to equalise the load carried by individual teeth, to make the relative depth of cut between adjacent roughing teeth more or less equal; but it is only the grinding of the finishing teeth which is really critical. If the amount of metal left to be removed by these teeth is reduced to a minimum, tool life will be prolonged and regrinding costs will be kept low.

Typical applications of broaching are spline and keyway cutting, rifling, and the machining of non-circular holes. The workpiece is clamped to the machine table, and the broach is either pulled or pushed past it. The "pull" method is most frequently employed, for with it heavier cuts can be made and longer broaches used; but in finishing operations and for light cuts, the "push" method is to be preferred. It can be carried out on a simpler machine, which allows some reduction in cycle time.

8.19 Dies. The ordinary round die, commonly known as the *button die*, is still widely used even though its future is seriously threatened by the development of the releasing die head holding chasers which can be removed for regrinding and replacing when worn. It must therefore be briefly considered here. An example is illustrated in Fig. 8.22 below.

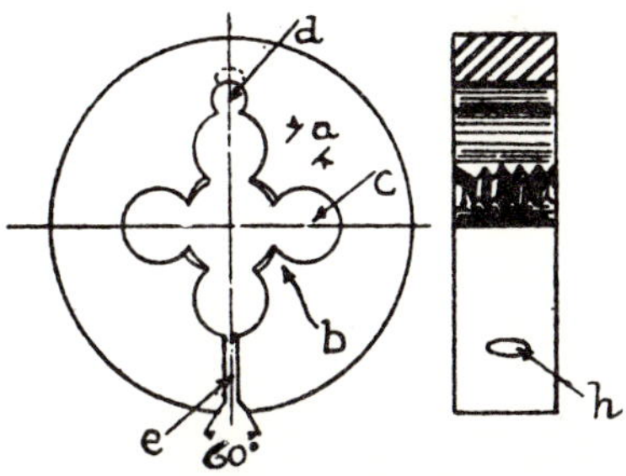

Fig. 8.22 A Button Die

In deciding the proportion of the teeth *b* to the size of the clearance holes *c*, a compromise must be effected between the two. Most dies are made with four teeth, so a reasonable diameter for the clearance holes will be about four-fifths of the threading size of the die. The land *a* (in other words, the width of the teeth) must not be too large lest it generate friction which could cause pieces of the thread to be twisted off in the die. It will be noticed that both the size and the position of the clearance holes affect the width of the teeth, and also the amount of rake (if any is required) on the cutting edges.

The centre of the clearance hole should be so placed that every clearance circle cuts away about one-eighth of the circumference of the centre hole.

The small hole *d* is inserted in order to add spring and so facilitate the adjustment of the die. The hole is sometimes elongated, as shown in dotted outline, but it should in no case extend further than half-way between the clearance hole and the outside of the die.

The slot *e* is usually made about 1 mm wide, and its front is opened at an angle of about 60° to enable the tapered adjusting screw in the die holder or adapter to engage in it. The three countersunk depressions *h* are arranged 120° apart, one at the back and one on either side of the slot. They are for adjusting purposes, and their diameter should be made to suit the diameter of the screws in any particular holder in use. It should nevertheless be at least one-third of the thickness of the die.

Sometimes slot *e* is omitted, and the button die is left completely solid.

8.20 Hinged Button Dies. The great disadvantage of button dies is that they cannot be easily sharpened, and that their cutting angle, rake and so on cannot be varied to suit the particular type of material to be cut. Provided with a hinge as shown in Fig. 8.23, however, they can both be re-ground and given any degree of rake required—the only disadvantage being that in this form they cost more to make.

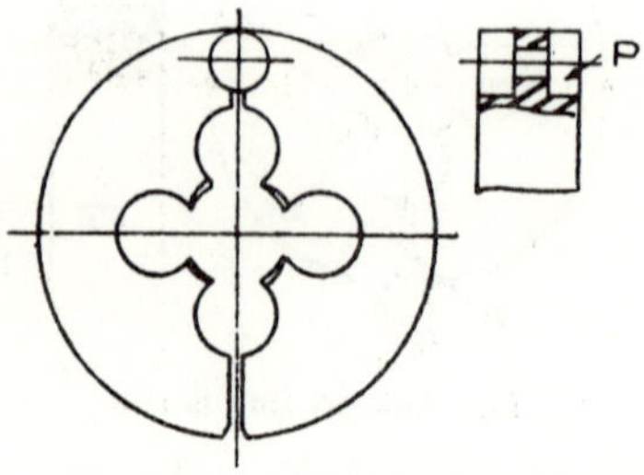

Fig. 8.23 A Hinged Button Die

If T be the thickness of the die, the diameter of the head of the hinge-pin **P** should be about T/2, the thickness of the hinge-pin itself and the length of the pin between the heads about T/3, and the diameter of the pin T/4.

Dies should be made of special-purpose steel, any alteration in the form or pitch of the thread after heat treatment rendering the die useless. If either the size or the form of the thread is to be in any way unusual, the designer should indicate on his drawing that suitable identification marks are to be stamped on the die before it is finished.

8.21 Spring Dies. This type of die, illustrated in Fig. 8.24, is more used in the United States than in Great Britain, and the writer is not aware of any particular virtue it possesses over the ordinary button die.

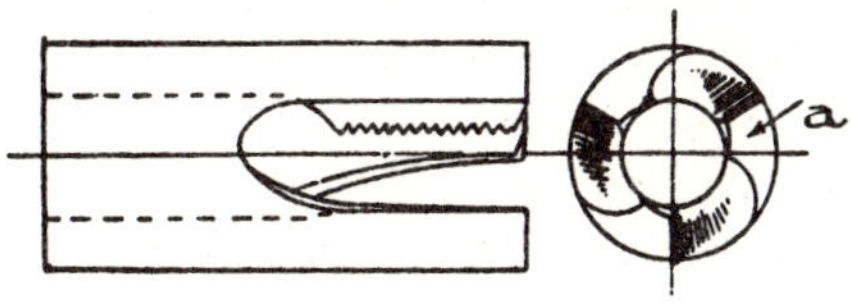

Fig. 8.24 A Spring Die

The gashes which form the cutting edges, usually three in number, are made with a milling cutter having an angle of between 60° and 70°. The length of the threaded portion is from one-and-a-half times to twice the diameter of the threaded bore. The lands *a* are proportioned as for button dies in that their width should be about one-eighth of their threaded bore.

The finished die should be hardened and tempered so as to provide spring for the jaws. This is necessary because adjustment is obtained by slipping a collar over the jaws, the latter having three grub screws by which they are set in or released.

If the die is too hard, however, the jaws will break on adjustment. If it is too soft, the jaws will not open out again after having been once set in. The degree of hardening applied is therefore critical.

8.22 Taps. Taps for ordinary commercial purposes are quite simple to make, but if either the form or the pitch of the thread need to be very accurate, the dimensions of the standard thread must be modified.

In the absence of a thread gauge, a threaded bore is sometimes checked with the aid of a cylindrical plug gauge whose diameter is approximately equal to the core diameter of the thread being produced. If the plug gauge enters, the hole is deemed unsatisfactory. When inspection on this point is strict, the correct size can be obtained by increasing the depth of the thread on the tap by a few hundredths of a millimetre.

When the form of the thread on a tap is distorted by final heat treatment to such an extent that the fit of the finished component is impaired, the effective diameter which governs the angle of the thread can be corrected until the fit becomes adequate.

Ordinary hand taps are rarely made in the user's shop, for they can be bought more cheaply from firms who specialise and produce them in very large quantities. Taps of special form of thread and diameter, however, are often drawn by the Jig and Tool Department. No dimensions can usefully be given for length or diameter of the shank of a tap, for they are often decided by the conditions of the job or by the nature of the holders available. Other factors which need to be considered include the nature of the material from which the tap is to be made, and the thickness of wall round the hole. Standard types of cutter for taps exist, and are identified by numbers. Convex milling cutters are also satisfactory in some cases, and a rough balance between the land and the flute can often be obtained by dropping such a cutter into the tap by a distance equal to one-quarter of its own diameter (see Fig. 8.25). Too much land causes friction and makes tapping difficult, but too little land reduces tap life. Maximum chip room is always essential.

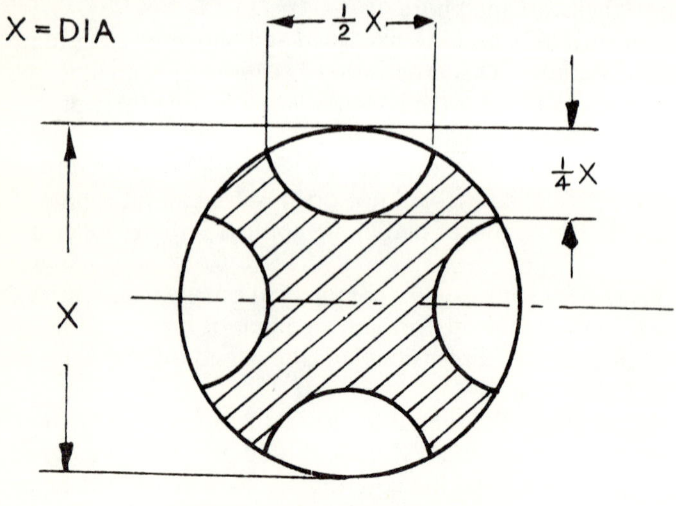

Fig. 8.25 Typical Dimensions for the Fluting of a Tap

It is now also possible to grind the thread of a tap instead of cutting it, and this method makes it possible to give the thread a measure of relief—a combination which makes for free cutting.

Taps used for tapping nuts are known as *nut taps* or *tapper taps.* They have a long shank and usually three of four flutes. But if the hole to be cut is long, and especially if it is to be cut in soft material, two flutes only are often preferable.

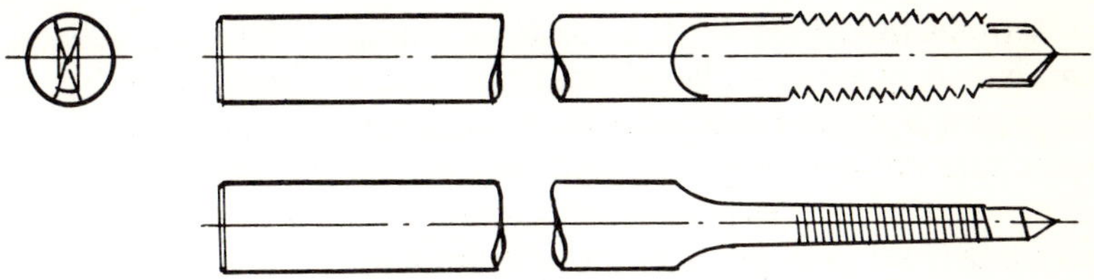

Fig. 8.26 A Combined Drill and Tap

8.23 Drill and Tap Combined. The type of drill and tap combined illustrated in Fig. 8.26 is often used in the production of cheap electrical parts. It is quickly made by threading a piece of silver steel, and then turning down a few of the front threads to core size and coning the point. Two flats and the cutting edges are ground, after which the tool is hardened and tempered.

CHAPTER 9

Press Tools

9.1 Introduction. Presswork consists of stamping out and shaping components made of malleable material such as sheet steel or brass. Irregular shapes and cut-outs like, for example, the body-panels of motor-cars are normally produced by a presswork operation.

The range of articles produced by means of the press is enormous, extending from pins, pen nibs and picture-hooks to steel coffins and radiator frames. Presswork can be used to save weight in a product without impairing its strength, or to increase the strength of a product without adding to its weight. A flat piece of 0·8 mm sheet steel can be easily bent in two hands, but just try and do the same thing with it after it has been well corrugated!

So many and various are the advantages of presswork, indeed, and so bright is its apparent future, that the reader is advised to make a special study of this interesting branch of engineering.

9.2 Simple Blanking Tools. The simplest possible type of blanking tool is illustrated in Fig. 9.1, in which the blank is a round disk.

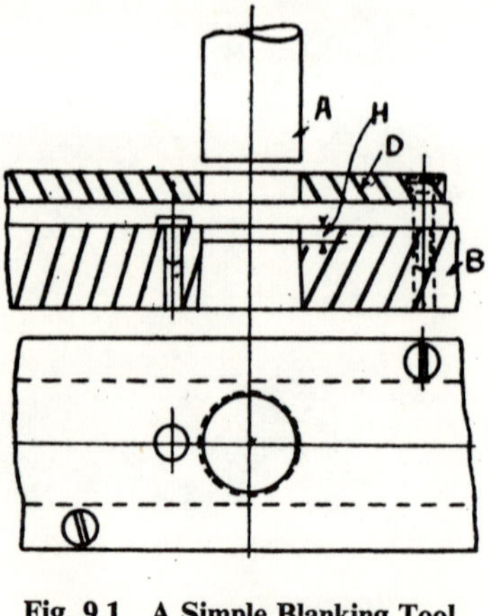

Fig. 9.1 A Simple Blanking Tool

The fit of the punch **A** in the die **B** depends on the thickness of the material being punched. If it is heavy and thick, greater clearance must be allowed between the punch and die so as to lessen the strain of punching.

If a punched *hole* is to be accurate to size, any clearance must be taken from the *die*. If it is necessary for the *blank* size to be accurate, the clearance must be taken from the *punch*. A useful general rule is to allow about one-tenth the thickness of the material.

The punch should also be a good fit in the stripper plate **D** in Fig. 9.1, for this makes it less likely that the tool setter will shear the edge of the punch or die when setting them in the press.

9.3 Multiple or Gang Tools. Two or more blanking punches can be mounted in a single holder, but in the arrangement of such punches two opposing dangers must be weighed in the balance. If the punches are set too far apart, much material in the sheet or strip will be wasted; but if they are spaced too close, the scrap from the strip will often prove difficult to handle. The general practice is to allow about 1·5 mm between adjoining punched holes, or (if it be greater) an amount equal to the thickness of the material being blanked. Note that the punches themselves cannot be placed as close together as that. They would be located as shown in Fig. 9.2. The first stroke of the press punches out two blanks **A**. The material is then fed along and those at **B** are completed (see right-hand half of Fig. 9.3).

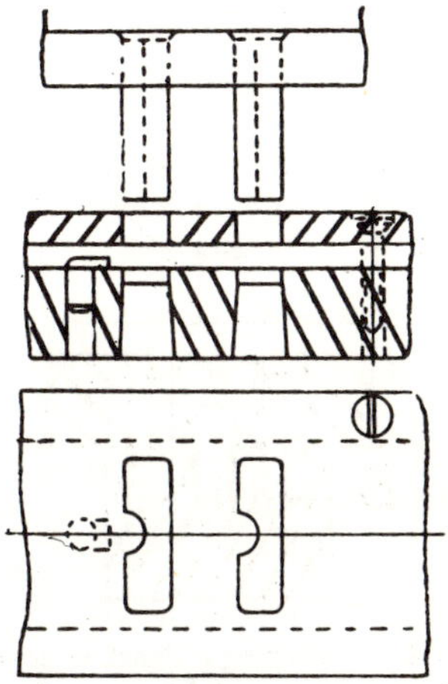

Fig. 9.2 A Multiple, or Gang, Tool

To avoid waste of metal when blanks of irregular shape are being punched, the strip of metal is often fed through the press blanking one side (**A** in Fig. 9.3) and then fed through again so that the opposite side **B** is punched.

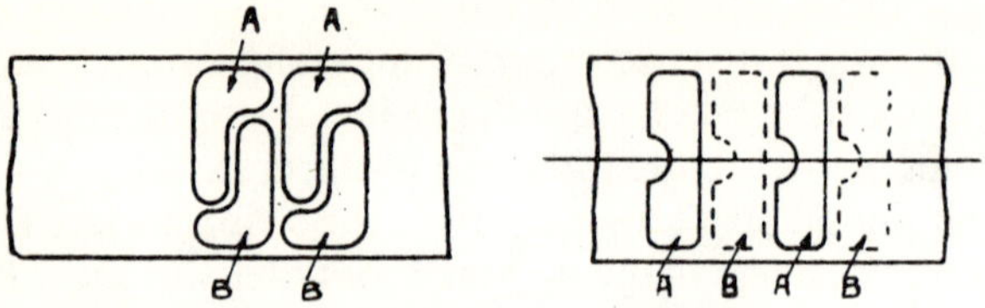

Fig. 9.3 Arrangement of Punches for Blanks of Irregular Shape

9.4 Punch Holders. The most usual method of holding punches is illustrated in Fig. 9.4A, the punch being made an accurate force fit in the punch plate and the back of the punch being then riveted over.

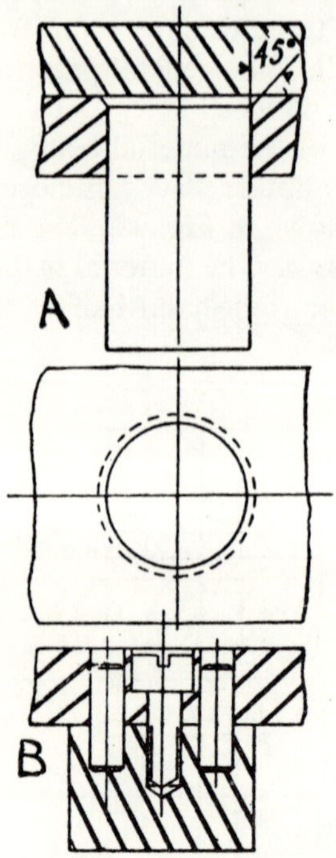

Fig. 9.4 *A*. **Holding a Punch by Riveting.**
B. **Holding a Punch with Screw and Dowel Pins**

Another method, used especially with bending and forming tools, is shown in the lower portion (B) of Fig. 9.4, where the punches are held by means of a countersunk headed screw and dowel pins, the latter passing through the back of the punch plate into the punch itself.

When fine punches that are liable to break are being used, provision must be made for quickly extracting any broken pieces. One way of doing this is illustrated at Fig. 9.5C, where the punch is held by an ordinary set screw; and another way in Fig. 9.5D, where the punch is made with a head and is passed through the punch plate in a very accurate fit until the head itself is sunk below the surface.

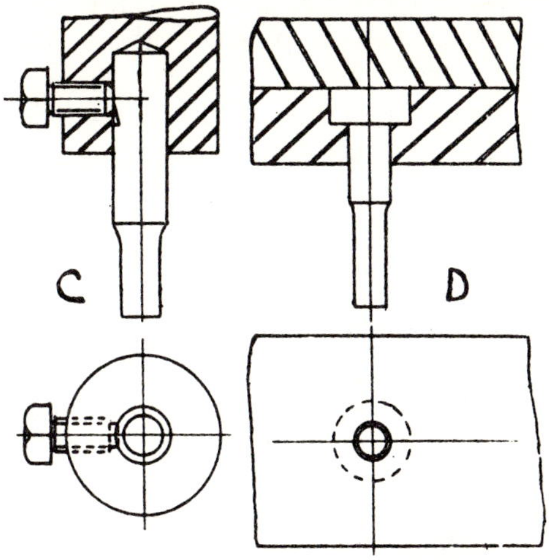

Fig. 9.5 Two Methods of Holding Punches so that they can be easily Extracted if Broken

9.5 Bolster. This is the name given to the cast-iron or steel bed plate on which the die is fixed when it is in the press. Some dies are fixed permanently on the bolster, while others are removable. The form of the bolster illustrated in Fig. 9.6 consists of an iron casting having

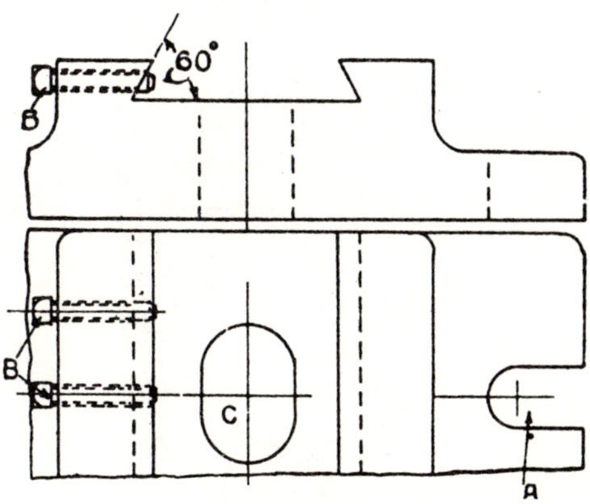

Fig. 9.6 A Bolster

two lugs **A** for holding-down purposes, and a dovetail seating into which fits the die. The set screws **B** engage in small depressions in the side of the die, so holding the latter securely.

The hole **C** must not be made large enough to weaken the bolster, yet there is a converse danger should it be made too small. If a blanking die were to be fixed on it and the hole were to prove not large enough to allow the blanks to pass through it as they fell, jamming would build up until either the punch or the bolster would finally break.

Bolsters of another pattern have no dovetail portion but are quite flat, the die being bolted direct on to the face.

9.6 Strippers. Once the punch has pierced the workpiece, it would normally carry the punched-out piece of metal with it on its return stroke. These unwanted pieces can only be removed with difficulty, and a stripping device of some sort has to be included in the equipment. The most usual method of stripping was seen in Fig. 9.1, where the plate **D** would automatically remove the metal from the punches as the latter passed upwards through it. This form of stripper has the further advantage of giving support to the punch —but also the grave disadvantage of making it practically impossible to see the punch at work.

A punch and stripper combined is shown in Fig. 9.7, where the punch **A** rides inside the stripper sleeve **B**. When it descends, the punch passes through the material and the sleeve is pushed up; but as it is withdrawn, the action of spring **C** on the plate and on the pegs **D** causes the sleeve to force the material off the punch.

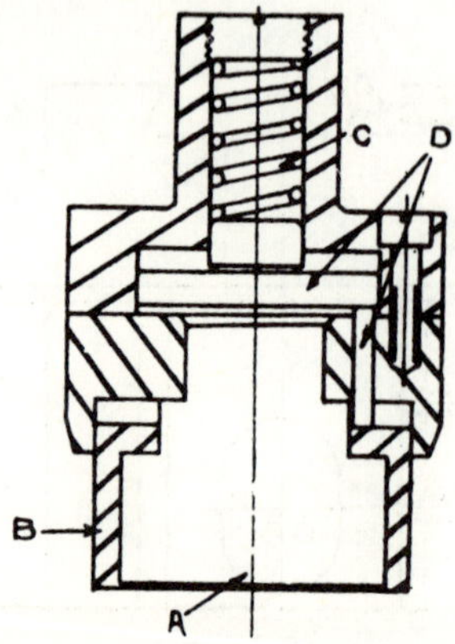

Fig. 9.7 A Punch and Stripper Combined

9.7 Stops. Stops are arranged on the die in such a way as to determine the relation of one blank to another. The usual method of arrangement is shown in Fig. 9.8, where a steel peg *a* is inserted as shown and the metal butts against it as it is being punched. The peg is bent as shown so as to make it possible to drill the hole *b*, in which it is to be held, further away from the die hole itself. This lessens the chances of the die cracking while it is hardening.

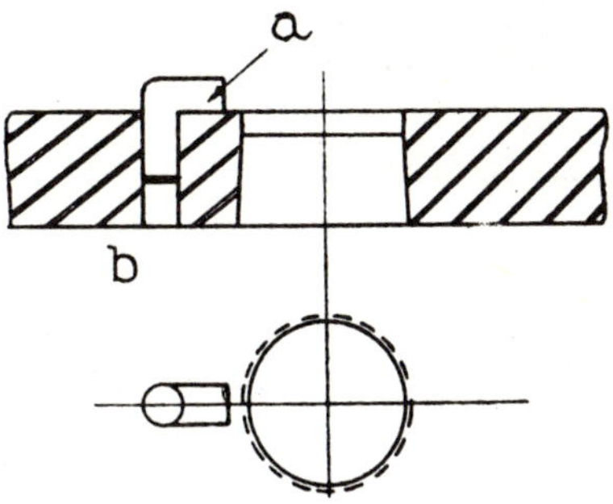

Fig. 9.8 A Fixed Stop

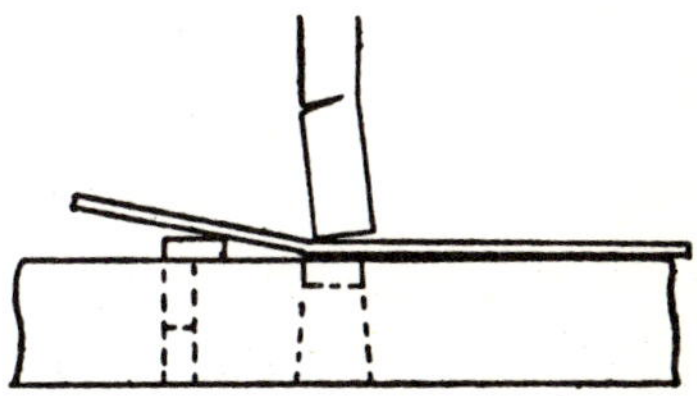

Fig. 9.9 A Disadvantage of the Fixed Stop

One disadvantage of this type of stop is that when the metal is being fed along it may, if not kept under proper control, ride over the stop instead of coming to rest against it. The result will be that the punch descends on to a sloping surface, as in Fig. 9.9, and may be damaged. This can be partly avoided by fitting a spring beneath the steel peg acting as stop pin, as shown in Fig. 9.10. If metal riding over the stop is now punched, the spring will be depressed and the metal will lie almost flat as the punch cuts through it.

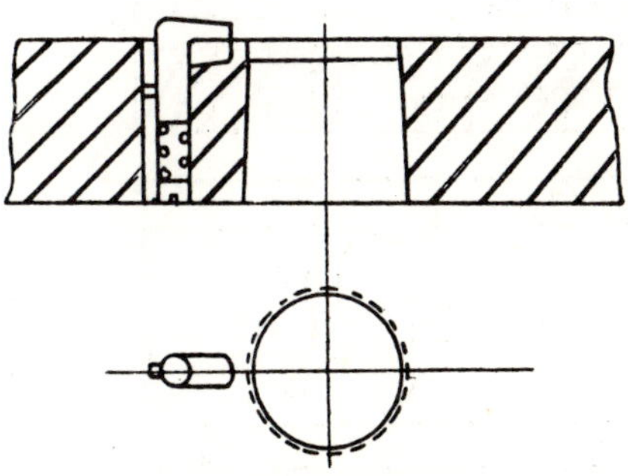

Fig. 9.10 Overcoming the Disadvantage with a Spring Stop

Fig. 9.11 illustrates another type of stop, in which the trigger *a* is hinged on the stripping plate *b*. As the metal is fed along to the left, the trigger lifts and drops into the most recently punched hole. The edge of the hole is then located against the trigger by giving the material a slight pull to the right.

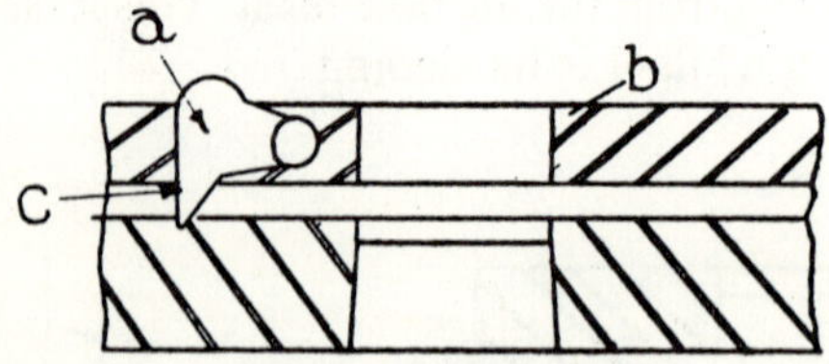

Fig. 9.11 A Trigger Stop

9.8 Follow Dies. These dies are so called because more than one operation is completed at each stroke of the press, the tools being located ("following") one behind the other. Considerable re-handling can often be saved in this way.

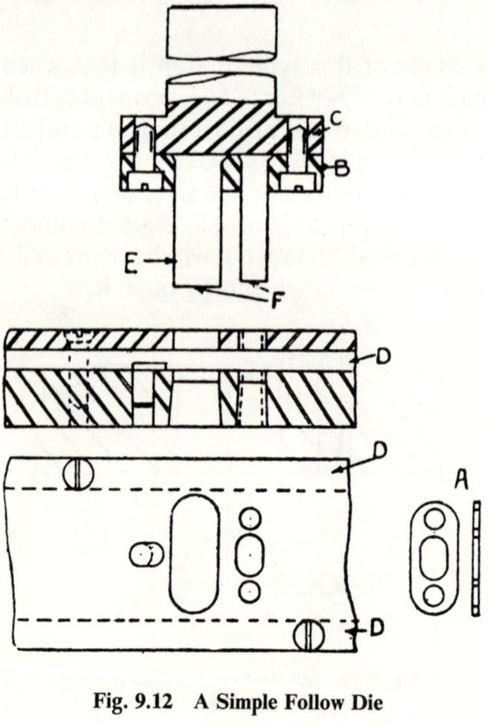

Fig. 9.12 A Simple Follow Die

Fig. 9.12 illustrates a simple tool of this type, in which every stroke of the press both punches out the slot and the two holes, and blanks out the shape of the component **A**. The punches are seen located in the punch plate **B** which is screwed to the holder **C**, the shank of the latter fitting the ram of the press.

The strips or guides **D** are made wide enough to permit free passage to the material being punched; but since the width of the latter may vary a little, with the result that the strip either acquires unwanted lateral movement or tends to stick between the guides, one of these is often fitted with springs which enable it to adjust itself sufficiently to cope with any variations in material width. The exact strength of the springs is something which has to be arrived at by a process of trial and error.

Note that the heavy blanking punch **E** in Fig. 9.12 is somewhat longer than are the finer punches. This allows **E** to enter the work first, so that the vibration or shock which could damage the more delicate punches will have had time to die down before they make contact with the strip being punched.

None of the punches should be made too long lest they buckle; but allowance must be made for the grinding effect which takes place on the faces **F**. Exact die thickness will depend on the thickness and area of the piece being blanked. It can vary between 10 mm and 25 mm.

In cases where a very accurate relation must be kept between a hole and the outside edge of a stamping, it is usual to fit a pilot pin on to the end of the punch. This pilot pin enters the punched holes before the blanking punch and guides the latter to accurate contact.

9.9 Drawing Tools. These are used in the production of cups, shells, metal boxes, etc. A simple form, known as a "push-through", is illustrated in Fig. 9.13.

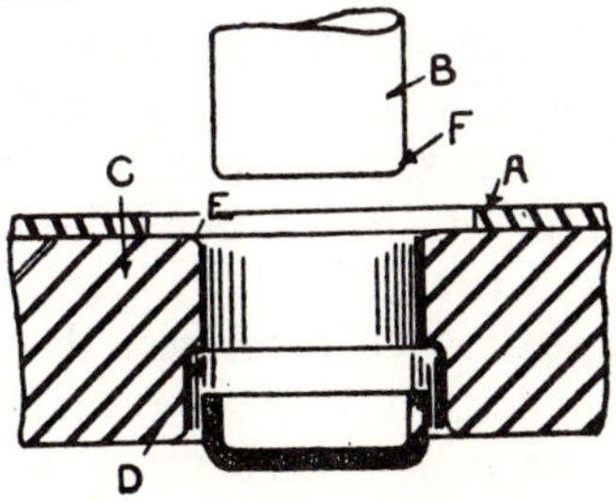

Fig. 9.13 A Simple "Push-Through" type of Drawing Tool

It is clearly only possible for a shell of straight formation to be drawn on a tool of this type, the blank for it being located in the guide plate **A**. The punch **B** as it descends forces the blank through the die **C**, and on its upward stroke strips off the shell as it catches on the edge **D**, which has been left sharp for the purpose. The edges of the die **E** and of the punch **F** must both have a radius appropriate for the operation, but sharp corners to the bottom of the component, if required, can only be given by a later operation.

The method of drawing described can only be applied to material at least 3 mm thick. Below this thickness, the edges of the metal entering the die tend to cockle and form a number of creases in the shell. To prevent this, a pressure ring as illustrated in Fig. 9.14 is employed, working as follows.

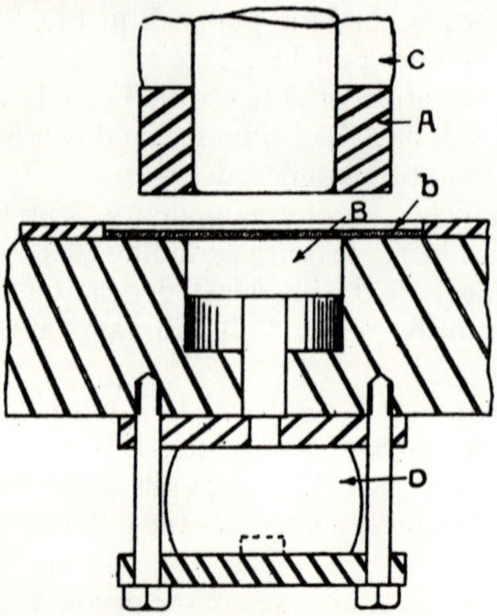

Fig. 9.14 Drawing with the aid of a Pressure Ring

The blank *b* is located in its guide over the die. When the punch descends, the steel ring **A** holds the blank firmly under pressure of the rubber buffer **C** for as long as the punch continues its downward movement. The blank is thus drawn beneath the pressure ring and the formation of creases is avoided. The ejector **B** operated by the buffer **D** removes the shell when it has been drawn.

The setting of the buffer **C** is critical. It must not be too strong, especially if the drawing punch has a small radius, lest it tear the metal instead of drawing it. Yet if its pressure is too weak, the danger of creases in the metal will re-appear.

Since the drawing operation stretches the metal, it is clearly impossible to state a size for the original blank, especially where the shell drawn is of irregular shape. But some idea of the size can be calculated by mensuration, and it is then a matter of experiment to find the exact shape of the blank. For *cylindrical* shells the diameter of the blank can be calculated roughly from the formula:

$$D = \sqrt{d^2 + 4dh}$$

where D = Diameter of blank,
d = Diameter of shell, and
h = Height of shell.

9.10 Re-drawing. When the depth of a shell is such that it cannot be drawn in one operation without risk of splitting, a re-drawing operation becomes necessary. Such an operation is illustrated in Fig. 9.15. It will be seen that the partly-drawn shell is located on a die having a "start" of about 45°. The punch then forces the shell through the die.

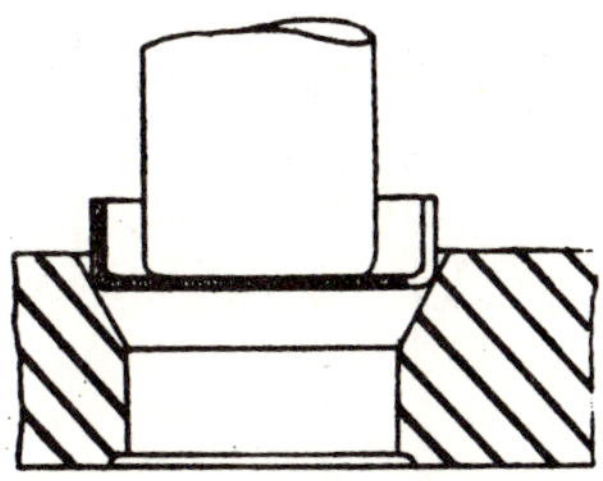

Fig. 9.15 A Re-Drawing Die

The number of draws necessary to complete a shell cannot be decided by rule. Among the variables which enter into the question are the quality of the material, the depth of draw required and the radius of the blank. The problem therefore remains one to be solved principally by experiment.

9.11 Combination Tools. If high production is to be achieved, the number of separate operations performed on a component must be kept to the minimum. Whenever possible, therefore, two or more operations are carried out with a single tool.

In Fig. 9.16, **A** is the die and **B** the blanking punch. **C** is the pressure ring supported by three studs equidistant from one another. These studs are screwed into the plate **D**, and the rubber buffer is compressed between **D** and the plate **E**. The drawing die **F**, which is of hardened steel, is located centrally in the die **A**. **C** should be a close fit between the blanking die and the drawing die. The "knockout" pad **H**, which also helps to form the shell, is a push fit in the punch **B**. A hole is usually located at **J** to allow air to escape behind the pad from **K**.

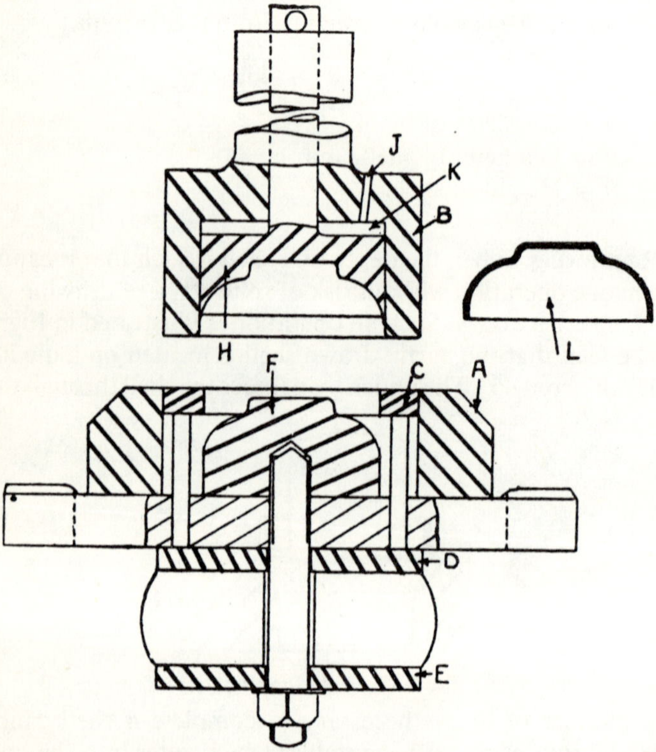

Fig. 9.16 A Typical Combination Tool

A single stroke of the press then both blanks and draws the shell, as shown at **L**. If the latter also needs embossing, the figures or letters to be embossed must be formed on the top of die **F** and on the underside of pad **H**—proud of one and recessed on the other.

9.12 Bending Tools. The simplest possible form of tool for achieving plain right-angle bends is shown in Fig. 9.17, where the strip **A** to be bent is located in a guide-plate **B** under the descending punch **C**.

The possibility of the piece slipping while being bent by this simple process is removed by employing a tool of the type shown in Fig. 9.18, where the strip of material to be bent rests on its guide and on the

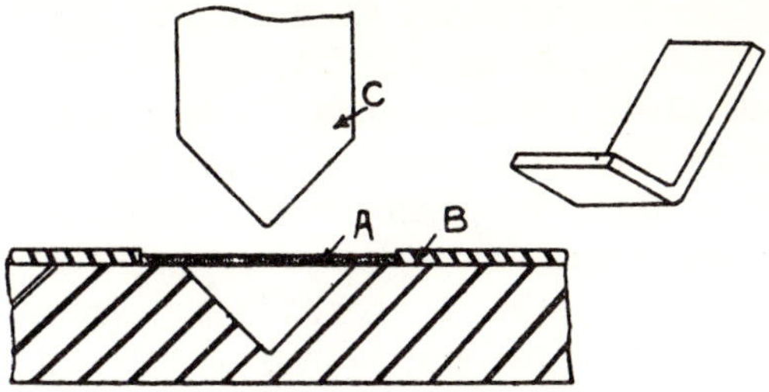

Fig. 9.17 A Simple Bending Tool

spring pad **A**, the latter descending with the punch and ensuring a perfect bend. To assist bending, a radius should always be arranged at *b*.

The width of the gap **B** should be made equal to the width of the finished component, plus 0·05 mm. The width of the punch **D** will therefore be the width of the finished component, minus twice the thickness of the metal, and minus 0·05 mm.

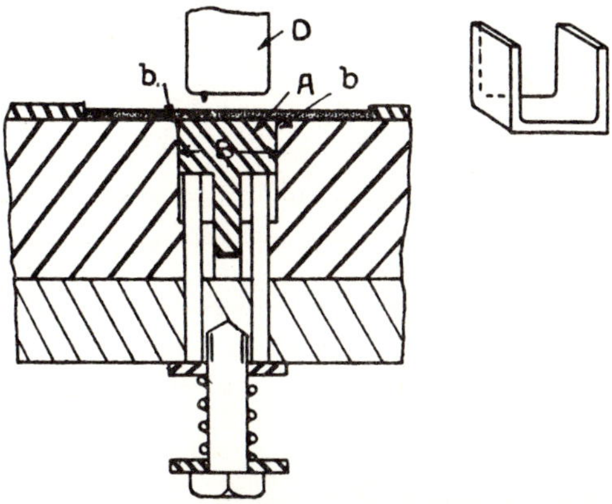

Fig. 9.18 A Bending Tool with Spring Pad

9.13 Materials. Press tools have the advantage over cutting tools that expensive high-speed steel is not used in their construction. The steel actually used depends largely on the types of material to be punched, but it would be safe to use carbon steel for all punches. For punching brass and even softer materials, however, a mild steel punch left unhardened is often used instead. Dies in all cases are made of carbon steel hardened and tempered. It is common practice to grind the hardened steel to the shape required, after the appropriate heat treatment has been given.

For piercing punches, ordinary silver steel or drill rod is good enough, but it should not be left too hard.

When a brittle plastic is to be punched or formed, as it often is in the making of electrical components, it can be prevented from cracking under treatment by placing a small water boiler near the press and running hot water through pipes arranged to pass through the die bolster. In hot weather it is generally sufficient to expose the plastic to the heat of the sun for an hour or so before it is used.

9.14 Built-Up Dies. The finishing of a one-piece die profile is often a tedious job. Grinding can sometimes, if only to a limited extent, be resorted to; but it is in this sphere that a built-up construction for the die has its advantages. It simplifies finishing to size and to accurate profile; it makes possible the replacement of a damaged part; and it is more economical of labour when the die is very bulky.

Fig. 9.19 illustrates a built-up die having four sections—all of them made to "nest" in the bolster **A** so as to counteract the side thrust which will be experienced when the die comes into operation.

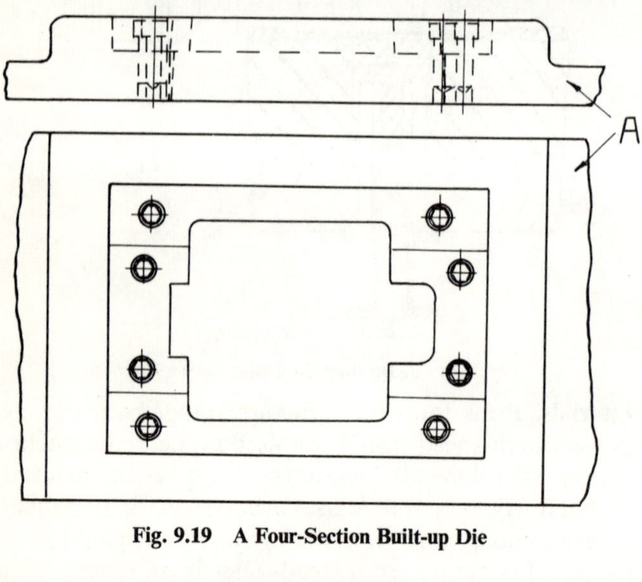

Fig. 9.19 A Four-Section Built-up Die

When the material to be blanked is thin, the die sections are often merely screwed or (less satisfactorily) dowelled on to the top of the bolster.

Built-up dies can usually be finished on a surface grinder, using formed wheels where necessary. By this method, the clearance angle is kept constant at about half a degree, and can be taken right up to the die face. The amount of clearance between punch and die increases when the face of the sections is later being ground, but the increase is so small that it can be neglected.

CHAPTER 10

Cam Design for an Automatic Lathe

10.1 Special-purpose Cams. Many of the different types of automatic lathe available on the market have their own built-in systems for controlling the turret and the various slide movements required; but in others this control is exercised by means of cams specially designed for each different component requiring to be machined. These special-purpose cams are not cheap to produce, and setting-up time for a given job will often be lengthy. But since such cams can be designed to optimise production rate and minimise idle time between operations *on a particular job*, cam-operated automatics are often capable of so low a cycle time that they become economic to use when a given component is required in large numbers.

The problems involved in the design of special-purpose cams form the subject of this Chapter; and it is proposed to take as an example the three cams which are needed to enable a Brown & Sharp automatic lathe No. "00" to turn out a quantity of mild steel screws of the dimensions shown in Fig. 10.1.

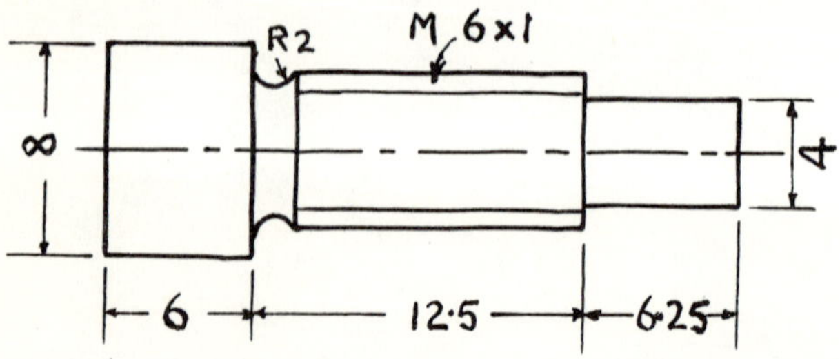

Fig. 10.1 Component to be Machined on Automatic Lathe

The three cams are as follows: (1) the lead cam illustrated in Fig. 10.2 which controls the movements of the turret holding the end-working tools; (2) the back cross-slide cam and (3) the front cross-slide cam, both of which are illustrated in Fig. 10.3. These control the movements of the cross-slides carrying the cutting-off and the form tools respectively.

Of the three, the lead cam is by far the most complex.

10.2 Feeds and Speeds. The first thing to be done is to work out the proportions of the several lobes of the cam. These in turn depend on the speed of revolution of the spindle, which must therefore be considered now.

For the machining of mild steel, a tool speed of some 25 metres per minute is about right. An examination of a surface speed table will give the number of revolutions per minute recommended for turning material 8 mm in diameter as 1,000. The number of possible speed changes on the machine are limited, however, and the nearest one to 1,000 is in fact 1,087 r.p.m. This will be accepted as "near enough".

For satisfactory results, the threading needs to be cut at a lower speed. About 10 metres per minute is considered correct. The nearest speed setting to this on the machine is 792 r.p.m.; and these two speeds, 1,087 and 792, will be accepted as giving good overall cutting results.

Another important factor in designing a cam of this type is the rate of travel, or feed, of the cutting tools. It is possible to design the proportions of the cam in such a way that the speed of the table can be varied to suit the requirements of any particular tool. For a component made of mild steel which does not need to be machined to very fine limits, a fairly coarse feed would be selected.

To machine the workpiece down to a thread diameter of 6 mm, only 1 mm needs to be removed from either side, and for this a feed of about 0·25 mm per revolution would not be too coarse. The further reduction of the thread diameter to form the 4 mm pin also calls for the removal of only 1 mm of material either side, so the same feed can be used for that job too.

For the cross-slide tools the feed cannot be so coarse as is adequate for an end-working tool. Selection of a suitable feed depends on a number of factors—among them the grade of material to be cut, its diameter, the width of the form tool and of the cutting-off tools, and the general condition of the machine itself. For the component under consideration, it will be assumed that the cut-off tool is of substantial width, the material nothing out of the ordinary and the machine bearings in good condition. Given that the form tool has only to make the recess under the head and to remove burrs, it can be taken that a feed of about 0·06 mm per revolution will not be too high.

10.3 Direction of Spindle Rotation. The threading die, as we have seen, must be run on to the work at the "slow" speed of 792 r.p.m.; but there is no reason why the high speed at which the spindle revolves when it is reversed for running off the die should not be put to good use. It is often arranged, therefore, that the reduction of stock diameter by hollow mill or box tool is done on the forward motion but the forming and cutting-off at a time when the spindle is running backward at high speed.

10.4 Total Revolutions per Piece. Once the required feeds and speeds have been decided, the operations to be performed on the workpiece can be identified and timed, the total number of revolutions of the spindle needed to perform them can be determined, and from them the time taken to machine a workpiece can be readily calculated.

For the machining of the component illustrated in Fig. 10.1, the operations required include turning down to 6 mm diameter as far as the head (which is best done with the aid of a hollow mill) and then turning the 4 mm nose portion (using for the purpose a box tool). The piece is then threaded; and as soon as the die is withdrawn, the form and cut-off tools move forward and operate simultaneously.

As a first step, the turret is indexed into position and the spindle is reversed, both actions occurring simultaneously. On the particular machine under consideration, the time taken for this is half-a-second, so that with the spindle running at 792 r.p.m. the indexing will take $6\frac{1}{2}$ revolutions. To this it is usual to add an allowance of two or three extra revs—and in this case the obvious addition is $3\frac{1}{2}$, so making the number of revolutions for indexing **10**.

The turning operation comes next. The length requiring to be turned (18·75 mm) divided by the feed per revolution (0·25 mm) makes 75 revolutions—to which 0·25 mm should be added for clearance, making **76** revolutions.

Now the turret is again indexed to bring into position the box tool for turning the point, taking as before **10** revolutions. The length of the point, 6·25 mm, is divided by 0·25, making **27** revolutions required for the operation—after which the necessary re-indexing of the turret adds another **10** revolutions.

The number of revs required for the threading operation is obtained by dividing the length to be threaded (12·5 mm) by the pitch of the threads (1 mm). $12{\cdot}5 \div 1 = 12{\cdot}5$, to which is added 2·5 revolutions for clearance, making **15**.

To run the die off the screw on reversal of the spindle adds another **15**, and then come the cross slide tools. The cut-off tool is the one which moves the greater distance, so all calculations should be based on the distance travelled by it. From Fig. 10.1 it will be seen that this distance is 4 mm (half the diameter of the bar), plus 0·25 mm to allow for the small angle at the point of the tool, plus a further 0·25 mm for clearance—making in all 4·5 mm. Dividing by the feed per revolution, which for the cross slide is 0·06, gives 4·5 ÷ 0·06 = **75**, which is therefore the number of revolutions required for the working of the cross slides.

During this latter operation, however, the turret has indexed round to the stock-stop position, making three indexings. It then has to feed forward another length of metal. So to the totals obtained so far there must be added another half-a-second; and this, together with an appropriate allowance, accounts for 14 revolutions at the faster speed.

We thus have a grand total so far of 10 + 76 + 10 + 27 + 10 + 15 + 15 + 75 + 14 = **252** revolutions. But that is not all.

10.5 Correcting the Number of Revolutions for the Higher Speed. Two speeds are employed in making the screw, and all calculations when proportioning the cam are based on the fast speed. It is therefore necessary to increase the number of revolutions of the slow-speed operations by the ratio between the two speeds, which in this case is

$$\frac{1{,}087.}{792}$$

This ratio is applied to the number of revs taken for all slow-speed operations. As an example, the slow-speed operation of indexing the turret, which it has been seen takes 10 revs, must now be amended to

$\frac{1{,}087 \times 10}{792}$ = **14** revs approximately—and all other slow-speed operations must be corrected accordingly.

The table overleaf (whose right-hand column "Percentage of Cam Surface" will be explained in Section 10.7) summarises the above calculations, and shows that the corrected number of revolutions taken for all the operations required is **307**.

10.6 Selecting Gears. Once the number of revolutions necessary to machine one piece has been determined, it is a simple matter to find the time per piece. With the spindle running at 1,087 r.p.m., it will turn 18 times in one second; so the 307 revolutions required per piece will be made in 17 seconds.

The machine designed to cut these cams was so built that the rate of motion of the cams controlling the various tools could be varied through the agency of a system of change gears. To machine the

ORDER OF OPERATIONS

	Revolutions	Corrected Revolutions	Percentage of Cam Surface
At 792 r.p.m.:			
Index turret and reverse spindle	10	14	5
Hollow mill	76	105	33
Index turret	10	14	5
Finish point (box tool)	27	37	12
Index turret	10	14	5
Thread on	15	21	7
At 1,087 r.p.m.:			
Run off die	15	15	5
Form. Index 3 times and cut off	75	75	23
Feed stock	14	14	5
	252	307	100

piece in 17 seconds it was therefore necessary to select gears having the correct number of teeth to give the required rate of motion—that is, to cause the cam-shaft to make one complete revolution in 17 seconds with spindle speed at 1,087 r.p.m. The relevant information is extracted from a table supplied by the manufacturers—and in this particular case, the gears required were found to be a 20-tooth wheel on the driving shaft and a 34-tooth wheel on the worm shaft.

10.7 Determining Cam Surface. To convert the number of revolutions already calculated into a measure of cam surface, a circle equal to the diameter of the cam blank is drawn and its circumference is divided into 100 equal parts. These hundred divisions represent, in the present case, 307 revolutions of the machine spindle; so that one division equals 307 ÷ 100 = 3·07 revs. The number of divisions, so calculated, which make up a given operation is listed in the third column of the table "Order of Operations" on the last page, under the column headed *Percentage of Cam Surface*.

10.8 Plotting the Cam. The cam blanks are provided with a central hole machined to fit the shaft on which they are to be located when in use, and with another hole 6 mm diameter which both provides a positive drive for the cam and also fixes its position on the shaft.

Both of these holes and the zero point of the 100 divisions are on the same radial line, and it is from this line that the spacing of the cam commences. Taking the operations one by one as listed in the "Order of Operations" table, the indexing and reversing take 5 hundredths, so that a radial line is drawn at mark 5.

The turning operation with the hollow mill, which takes 33 divisions, is then measured off and the next radial line is drawn at point 38. Since it is the job of the cam to take the end-working tools to their work and then return them, the length of the tool and its holder will clearly affect the position of the highest point of the cam in relation to the outer edge of the cam blank. (The latter is 100 mm in diameter for the operation under discussion, and it is assumed that standard equipment is being used on the machine.) The position of the highest point on the cam is found by measuring off from the point on the component where any particular tool finishes its cut, back to the turret face at the limit of its travel. If the length of the tool exceeds this measurement, it means that the highest point of the cam for the tool must be that amount from the outer edge of the blank, so shortening by the required amount the movement of the turret slide.

In the particular machine under discussion, the extreme forward position of the turret face is 45 mm from the spindle nose. The component being 24·75 mm long (Fig. 10.1), a distance of about 15 mm can be allowed between the turret face and the component, leaving 5 mm of bar projecting from the spindle for the cutting-off operation.

As regards the hollow mill, the measurement from the end of its cut to the turret face is $18{\cdot}75 + 15 = 33{\cdot}75$ mm. This means that, if the length of the hollow mill in its holder is (as it well might be) 40 mm, it will extend some 6 mm beyond the turret face, so making the distance between turret face and component exactly 15·25 mm.

The highest part of the cam on the lobe for turning the threaded portion will therefore be 6 mm from the outside measured on the radial line number 38. On line 5 a point is taken radially inwards from the rim by an amount equal to $6 + 18{\cdot}75 = 24{\cdot}75$ mm (that is, the distance from the outside edge already fixed plus the length of stroke called for by the component); and from this point a curve is drawn having a uniform rise up to line 38.

One way of drawing this curve accurately is to divide the space in the quadrant between lines 5 and 38 into an equal number of parts, and to divide line 5, beginning at the start of the curve and ending at a point 6 mm from the outer edge of the cam, into an equal number of parts. Using the cam centre as pivot, describe arcs through each of these latter points. The radial lines dividing the quadrant cut these arcs, and the lobe for operating the hollow mill is drawn through the points of intersection. The method is illustrated in Fig. 10.2.

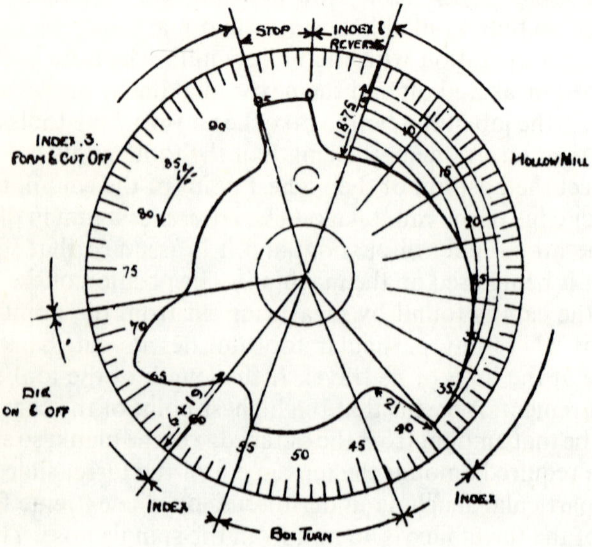

Fig. 10.2 The Lead Cam

Reverting now to the "Order of Operations" table, the re-indexing of the turret again occupies five divisions, and another line is drawn at 43. From line 38, the turret needs both to rotate and to draw back to accommodate the box tool; and to achieve this the roller on the actuating lever must drop from the high point it reached on 38 to a position which will retract the turret into position to operate the box tool so that it can finish the point of the screw.

The line drawn for this drop is normally made tangential to the centre hole of the cam; and at the bottom of the drop, an arc is described equal in radius to that of the roller on the cam lever. This arc then merges into the rising curve operating the box tool at the point where the curve begins. This point, in turn, is determined by the arc-and-radial method already described, the length of the tool this time being 36·5 mm, which passes the turret face by 15 mm. The curve for the box tool will, therefore, begin on line 43 at a point 15 mm + 6·25 mm from the outside rim of the cam blank, and will rise to a point 15 mm from the rim on line 55, with the 12 divisions required for the operation intervening.

Another drop is then made for re-indexing the turret, in the manner already described, extending five divisions from lines 55 to 60.

The lobe required to control the die comes next; but a method different from that used for ordinary cutting tools is needed to determine the extent of its rise. The method generally used is to multiply the number of corrected revolutions required for the threading operation (21) by the pitch of the threads to be cut (1 mm). Since it is essential to the production of good threads that the die should be neither forced on to nor torn from the work, an amount between 7 per cent and 12 per cent of this figure is then deducted. The full rise required is thus 21, less say 10 per cent, or about 19 mm.

When the lobe has been drawn, its high point is then removed for about 2·5 mm to enable the die holder to draw out of its socket and to finish up close to the screw head without the positive thrust of the cam.

The room for the die holder must now be provided as before, and it will be found that 6 mm must be added to the 19 at the starting-point of the lobe, the latter thus rising to within 6 mm of the rim of the cam blank. The curve for feeding the die to the work fills the space between lines 60 and 67, and for removing the die from the work the space between lines 67 and 72.

Note that for threaded work of high accuracy, the lobe needs to be more accurately drawn than this, since it is necessary to take also into account the different speeds at which the die is run on to and off the work, the relative position of the cam lever, and the pitch of the threads. For ordinary work, however, the method described will suffice.

After the threading operation the turret is re-indexed three times, during which period the form and cut-off tools on the cross slide are brought into operation. To achieve this, the shape of the cam after the threading lobe from line 72 consists of a portion of a circle of which the centre of the cam is the centre and whose radius of 30 mm continues for 23 divisions (see "Order of Operations" table again), so ending at line 95.

The last five spaces on the cam are occupied by the lobe required to feed forward the stock into the machine. The top of this lobe is an arc concentric with the outside of the cam forming what is known as a "dwell". This is because, since it is concentric and so has no "rise", the tool it controls remains stationary during its passage. The stop whose job it is to control the amount of metal to be fed forward is actuated by this lobe; the distance from its top to the rim of the cam blank is determined in the same way as before. So if the stop extends (say) 25 mm from the end of the component, it clears the turret face by 9·75 mm; and this is the distance the arc must be described from the outer edge.

Lastly, from the zero point on the face of the cam blank a line is drawn tangential to the centre hole, merging into an arc corresponding to the radius of the cam lever roll, which in turn merges into the turning lobe beginning at line 5.

The list of revolutions and gears required are then printed on the cam for the benefit of the machine setter, and the lead cam is finished.

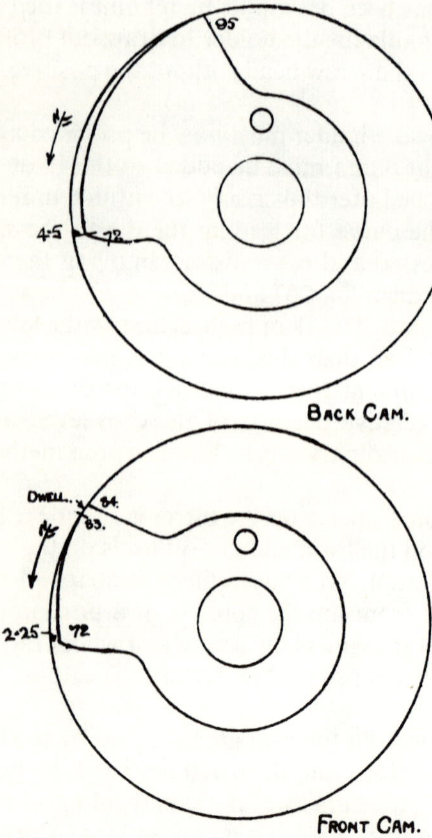

Fig. 10.3 The Back and Front Cross-slide Cams

10.9 The Back Cross-slide Cam. In comparison with the lead cam, both the back and front cross-slide cams are of simple design. Only one large lobe is needed on the back slide cam; and since the cross-slide operates at division 72 on the lead cam just completed and extends for 23 divisions, these are the radial limits of the lobe required. From the calculation of the numbers of revolutions necessary, it is known that the travel of the cut-off tool is 4·5 mm. The low point of the cam on line 72 is therefore 4·5 mm from the outer edge of the cam, as is shown in Fig. 10.3.

The remainder of the cam surface is drawn at a reduced radius of 30 mm from the cam centre. Lines 72 and 93 are tangential to the centre hole, but the former is given a slight radius as indicated. Both merge into the remaining portion of the cam.

10.10 Front Cross-slide Cam. Since the form and the cut-off tools operate simultaneously, line 72 is again used as starting-point. The extent of the cam area is found by calculating the number of hundredths as before.

The travel of the form tool is equal to the difference between the radius of the screw head and the radius of the undercut portion, plus about 0·25 mm for clearance. It is thus 2·0 + 0·25—from which the number of revolutions and then the number of divisions can be calculated. The low point of the cam will be found to occur on line 72, 2·25 mm from the rim, and to extend to line 84.

The last division on the front cross-slide cam is generally left unmachined and concentric, so providing a "dwell" for the form tool which enables it to remove burrs thrown up by the cut-off tool.

CHAPTER 11

Gauges

11.1 Definition and Purpose. There are many different shapes and sizes of gauge, but they can all be defined as accurately dimensioned pieces of metal used to save time and the employment of unnecessarily highly skilled labour in checking the dimensions of workpieces or of other less precisely made gauges. When repetition work has to be produced quickly and accurately to dimensions, for example, it is essential to obviate the need for taking repeated measurements with precision instruments such as micrometers and verniers—especially if the measuring will have to be done by unskilled or semi-skilled labour. Something accurate enough, but much quicker to use, is required.

Another important use for gauges is in securing the interchangeability of machine parts at the lowest possible cost. This is done by setting limits of accuracy within which the particular piece undergoing manufacture must fall, so allowing any component whose dimensions when finished fall within these limits to be made cheaply and to be assembled without difficulty into a more complex piece of equipment.

Take, for instance, the plug gauge illustrated in Fig. 11.1. Say that a hole has to be drilled or bored in a large number of components whose diameter, it has been determined, will serve the purpose if it falls within the limits of 20 and 20·25 mm. The plug gauge has one of its ends made to an accurate diameter of 20 mm, and the other end to an accurate diameter of 20·25 mm. If the hole produced allows the 20 mm end to enter, but not the 20·25 mm end—a test which can be quickly applied by anybody—its diameter must of necessity fall within the limits of acceptability.

The difference of 0·25 mm between its two ends is called the *tolerance* of the plug gauge.

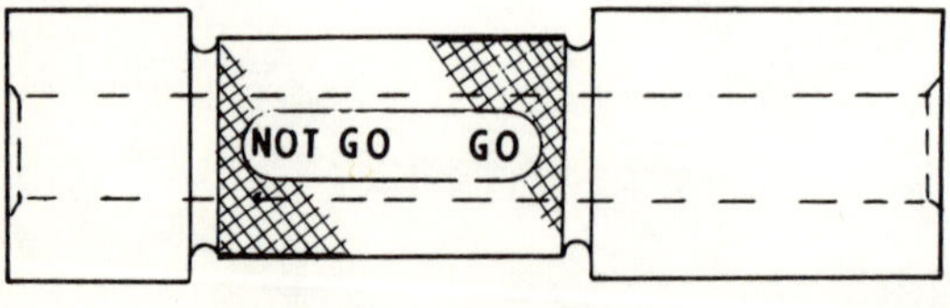

Fig. 11.1 A Simple Plug Gauge

11.2 Limits. The question of limits and the methods of fixing them is the subject of much controversy, and almost every firm practises a different system. Generally, the fixing of correct limits and the choice of gauging methods is made by the chief inspector in consultation with the designer of the article being produced.

Since in the building of jigs and fixtures, however, some system of limits is necessary to secure interchangeability and the correct fit, it is recommended that choice be limited to the wide range of grades, or *tolerance limits*, laid down in BS 1916. Depending on the class of fit required between mating parts, it will be possible to choose from this British Standard publication suitable grades for both the "shaft" member and the "hole" member of the jig or fixture required.

11.3 Fits. When a particular pair of components is assembled, the fit between them will be either a *clearance fit* (when the shaft is smaller than the hole it is to go into), or an *interference fit* (when the shaft is larger than the hole). When the tolerances allowed are such that shafts selected at random from a large batch enter some holes but not others, the class of fit is called *transitional.*

Common workshop terms used to describe classes of fit lying between clearance and interference are *running fit*, *push fit*, and *drive fit*. The meaning of each is obvious, and examples of all of them can be selected from the Standard.

11.4 Plug Gauges. Fig. 11.1 showed a simple type of plug gauge. The body is knurled to facilitate handling, and provision is made on it for engraving a reference number and the words GO and either NO GO or NOT GO. The NOT GO end is made appreciably the shorter of the two so as to ensure quick identification.

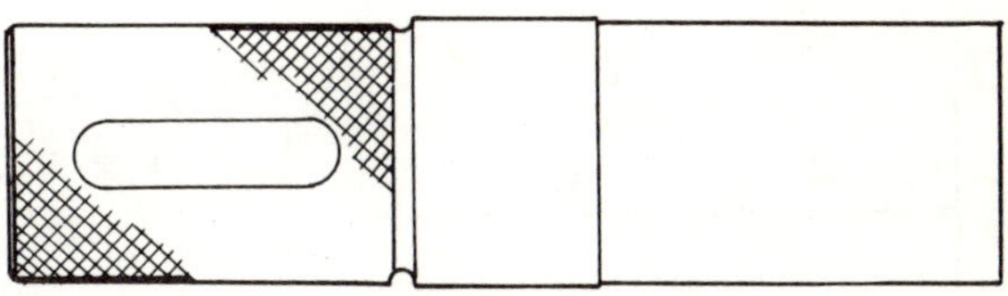

Fig. 11.2 Another Type of Plug Gauge

Another simple type of plug gauge having a limited application is illustrated in Fig. 11.2. In it the NOT GO portion is located immediately behind the GO portion, and the gauge can only be used with safety if the GO end can pass right through the work. Concentricity must obviously be perfect, but as a time saver the type has advantages.

Heavy or bulky gauges are difficult to handle with accuracy, and should be avoided. If they are over 50 mm or so in diameter, they can be lightened by having drilled through them a hole with a diameter of about half that of the gauge itself. Or the two end-pieces can be made separate and hollow, as shown in Fig. 11.3, so that they can be replaced when worn without scrapping the entire gauge.

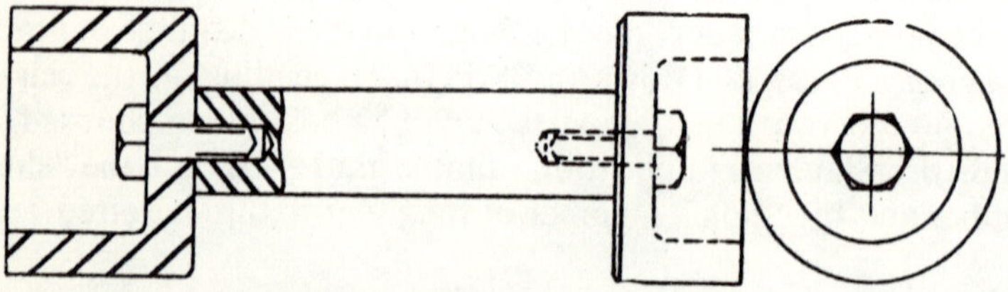

Fig. 11.3 A Plug Gauge with Hollow Replaceable Ends

Sometimes, if the hole to be gauged exceeds 80 mm in diameter, a plate gauge of the form shown at Fig. 11.4 is used. When the limit on so large a hole is close, it is difficult to hold a heavy gauge sufficiently square for it to be able to enter the hole. A light gauge which is at the same time robust enough not to be easily distorted is required, both to save time and to achieve accuracy. The gauge shown in Fig. 11.4 would typically be made from sheet steel about 10 mm thick. The diameter is removed or relieved as indicated at **C**, leaving a land at **A** which makes it necessary for the inspector to test any hole in two positions to ensure that it is perfectly round. The handle **B** is often made of steel tube for lightness, and the plate can also be holed as at **D** to achieve the same purpose. A blank ground on the holder will have the number of the gauge and the limits it will indicate marked on it.

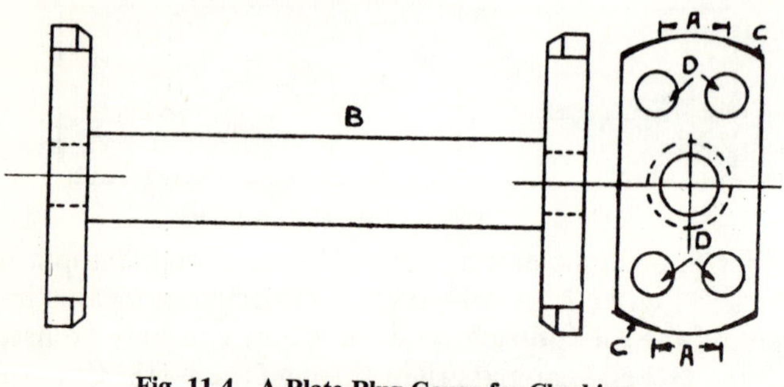

Fig. 11.4 A Plate Plug Gauge for Checking Large-diameter Bores

For very fine holes of about 3 mm diameter or less, a useful and cheap form of gauge is illustrated in Fig. 11.5. It consists of two pieces of silver steel wire linked together, one on the high and the other on the low limit, each having a steel or celluloid disk attached marked with the details.

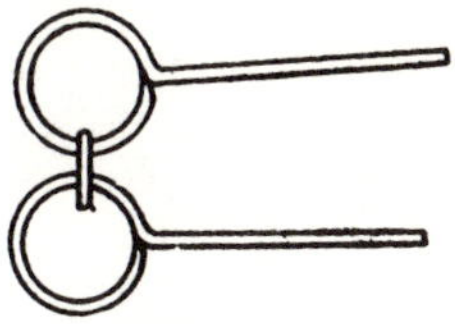

Fig. 11.5 Pin Gauges for Checking Small-diameter Holes

11.5 "Pilot" Plug Gauges. All mechanics understand the care that is needed to insert a plug gauge properly into a hole, particularly when the limit is very fine. The dangers are that the gauge may become wedged, that the work may be damaged or loosened in the machine chuck, and that much time will in one way or another be wasted. A simple device has been developed to minimise these dangers, consisting of a vee groove machined round the gauge itself leaving a land 3 mm from the bottom whose diameter is slightly reduced so that it acts as a pilot to the gauge itself. With this device, the gauge can be entered in perfect alignment without difficulty, the lip left by the groove providing a lead.

In a blind hole, however, the use of such a gauge would not be satisfactory because the diameter at the bottom of the hole, for a distance equal to the groove width plus the pilot length, could not be checked. A modified type of gauge known as the "Semi-Pilot" is therefore used instead. In it the groove does not extend completely round the gauge, but two portions of the full diameter situated diametrically opposite one another are left blank. When such a gauge is being used, note that it must enter square with the hole lest the gauge jams as would an ordinary plug.

Note also that the semi-pilot gauge is the subject of a British patent, and that a licence is necessary before it can be used.

11.6 Calliper Gauges. Often known as *gap gauges* or *snap gauges*, gauges of the calliper type are used for measuring external diameters and thicknesses. An example is shown in Fig. 11.6.

Since there is always a tendency for the jaws of a calliper gauge to spring slightly open, part **A** in Fig. 11.6 must be strong enough and of sufficiently broad section to withstand it. The thickness **B** will depend on the article being checked. If it has a shoulder, as at **C** in Fig. 11.7, a thin section of about 6 mm will suffice. The stem **D**, however, would call for a gauge of wider section, less for strengthening purposes than to enable the inspector to check over the length of spindle more quickly.

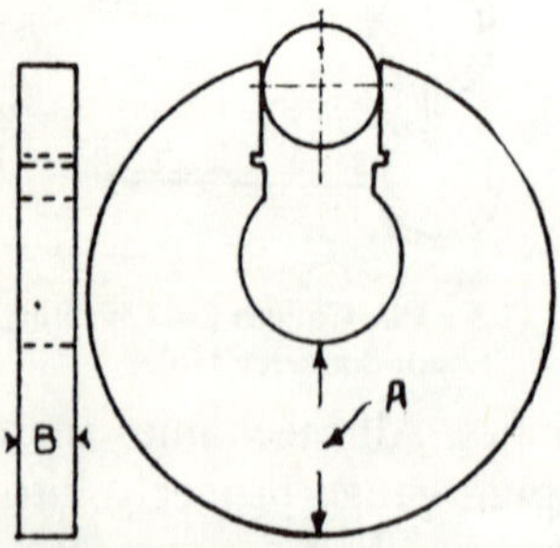

Fig. 11.6 A Typical Calliper Gauge

It is always best to err on the wide side with all gauges of the calliper type so as to secure a good seating, as at A in Fig. 11.7. Compare the small area of seat at B, where there is obvious risk of the operator not holding the gauge square and so obtaining a false result.

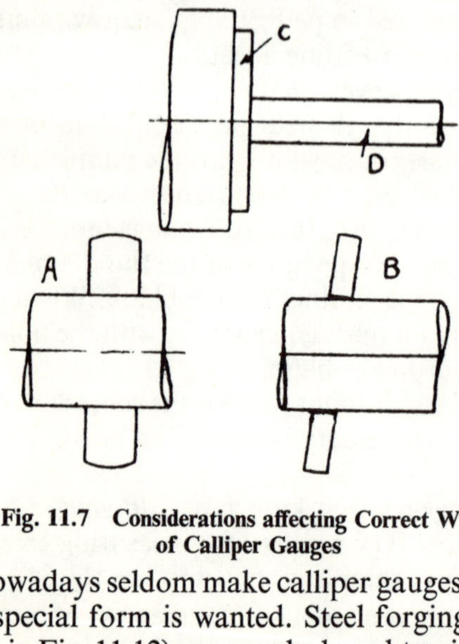

Fig. 11.7 Considerations affecting Correct Width of Calliper Gauges

Designers nowadays seldom make calliper gauges of sheel material unless a very special form is wanted. Steel forgings of very strong section (as at A in Fig. 11.12) can now be bought quite cheaply in lieu.

11.7 Scissor Gauges. To check the diameters of recesses machined into a component, a special form of calliper gauge known as a *scissor gauge* is used.

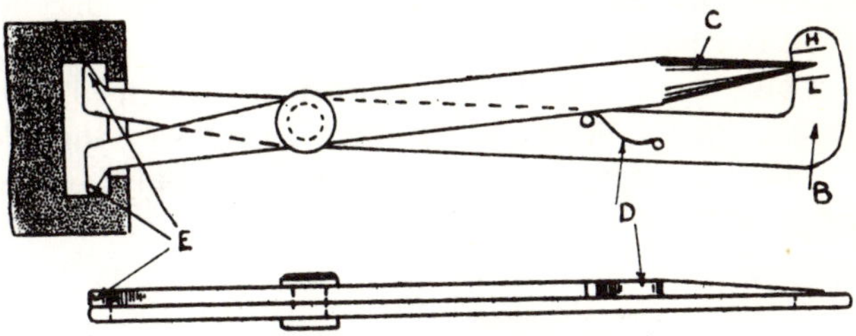

Fig. 11.8 A Scissor Gauge

Fig. 11.8 shows a scissor gauge in position. It will be seen that the two limbs are pivoted at a point which is not too near the measuring points lest the gauge become clumsy when held in one hand, but which is yet far enough away from these points to ensure that the limit allowed in the recess is adequately multiplied on the dial B. Obviously, the point of the limb C must rest between the high and low limits marked on the dial when the component is correct. The spring D keeps the jaws E in contact with the walls of the recess being checked, and also makes it possible for the gauge to be used with one hand.

A specialised form of scissors is used for rough measurement of the walls of cylinders or of other pieces where a direct measurement cannot be taken. The pivot pin is in this case arranged to be equidistant from both ends, so that the actual measurement taken on one end is shown exactly on the other.

11.8 Screw Thread Gauges. Used for checking the accuracy of pieces threaded either internally or externally, these are among the most expensive gauges to produce, the final hardening process calling for much care. If a suitable grade of steel is used, however, the risk of spoiling is greatly reduced.

Ring gauges are used to check external threads (see lower half of Fig. 11.9), and should always be externally knurled to assist release of the gauge in case it should become fixed on the component.

The theoretically ideal method of checking a thread is to use an optical comparator in conjunction with a ring gauge for male threads, and with a threaded plug for tapped holes. The comparator projects an image of the thread form of the piece to be checked on to a screen which also carries an accurate image of the same thread magnified up to 80 times. The profile of the thread to be checked can be superimposed on that of the model thread by manual adjustment, and any defects in its form are clearly exposed. In production work, however, this method is seldom possible, and the usual method of inspection is to use a plug gauge for the core diameter of a tapped hole and a threaded plug for the fit; while for checking screws a threaded ring gauge is used for the fit, and a form of calliper gauge to check the effective diameter.

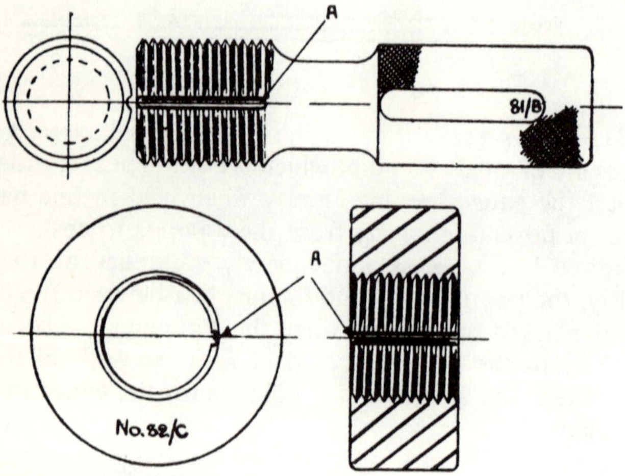

Fig. 11.9 Male and Female Screw Gauges

This latter diameter controls the flanks of a thread—the portion which takes the hard work when in use. The gauge for the effective diameter looks not unlike that illustrated at **E** in Fig. 11.12, which it will be seen provides a GO and a NOT GO. Gauges involving the three-wire principle are also in use, but require careful handling.

The design and making of gauges for threaded work forms a specialised class of tool work. In Fig. 11.9, for instance, the groove **A** is not a standard device but has been introduced to help clear the thread of any dust or congealed oil. Ring gauges, too, are sometimes split axially and given a tangential screw to provide a fine adjustment.

11.9 Profile Gauges. This type of gauge, as its name suggests, is used to check the shape of a component or part of a component. One such is shown at **C** in Fig. 11.12, and again in outline in Fig. 11.10 where it is used for checking the form and position of the grooves in the component **B**.

This gauge would be improved by mounting it on two small blocks so that its profiled edge was raised off the surface plate by an amount equal to the diameter of the component, as at **E**, thus facilitating accurate checking.

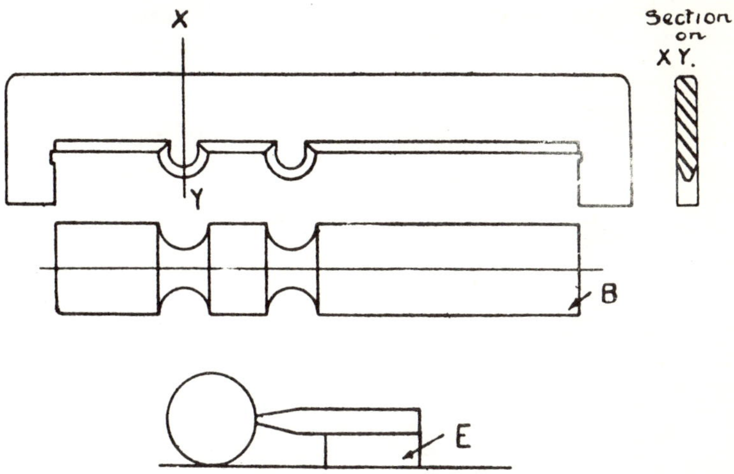

Fig. 11.10 A Typical Profile Gauge

Another type of profile gauge which is also used as a *position gauge* to check the relative positions of two different components is shown at **D** in Fig. 11.12.

11.10 Combination Gauges. Fig. 11.11 illustrates a type of gauge in which the functions of two or more of the gauges already described are combined. The component is an automobile engine valve, usually made from an alloy steel forging and ground all over. The gauge shown can be used to check the diameter and straightness of the stem **A**, the diameter of the seat **B** and its angle to, and concentricity with, the stem, and lastly the position of the slot **C**.

The gauge fails in only one respect. It cannot check the stem should its shape be oval rather than truly circular.

11.11 Variety. In addition to those already mentioned, Fig. 11.12 illustrates a number of other gauges of different sizes and shapes, designed to serve a number of different purposes. They represent only a few of the many and various types of gauges used in a modern engineering workshop.

Shown at **F**, for instance, is a plate bearing three pegs whose purpose is to test the smooth running and correct centering of three gear wheels. If the gears are rotated slowly on the pegs, any irregularity in the size or form of their teeth is immediately noticeable. The pegs can also be used to check the bore diameters of the gear wheels. The base of the fixture is of cast iron, the pegs of hardened steel.

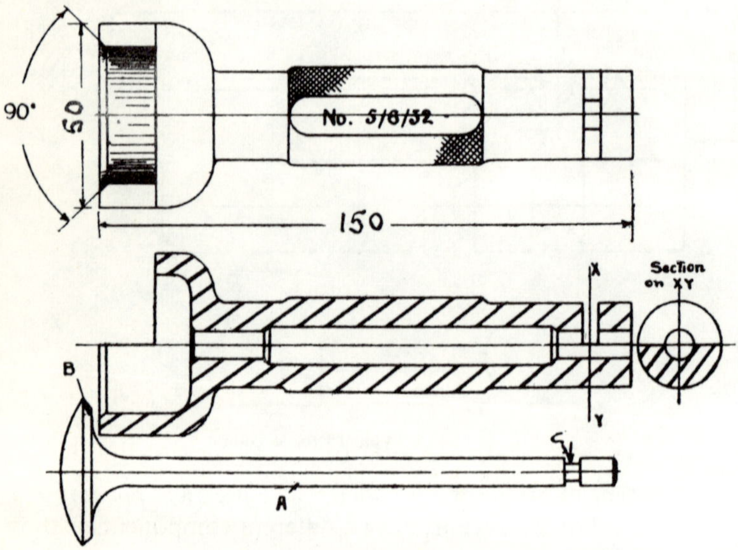

Fig. 11.11 A Combination Gauge, with Component to be Checked

The three gauges shown at **H** are used for checking the relative positioning to one another of the crank pins of a four-throw crankshaft. The thin stem is of steel tube with a knife-edged roller in hardened steel at **B**. The base is of aluminium, with a thin steel sole to lessen wear. This sole locates on the flange at the end of the crankshaft, and a gauge fitted to each crank-pin or journal ensures their correct relation.

J illustrates a plate length gauge made from 6 mm sheet steel, with the high and low limits projecting on either side.

The small calliper gauge at **K** is made from a piece cut from a bar of round section, and can be economically used for measuring small diameters up to about 5 mm.

The gauge **L** is used for checking the thickness of the rim of a gear wheel. The long upper edge is made to slide across the face of the work, thus eliminating any possibility of the gauge tilting and so giving a false result; while the high and low limits are stepped on the short arm of the gauge.

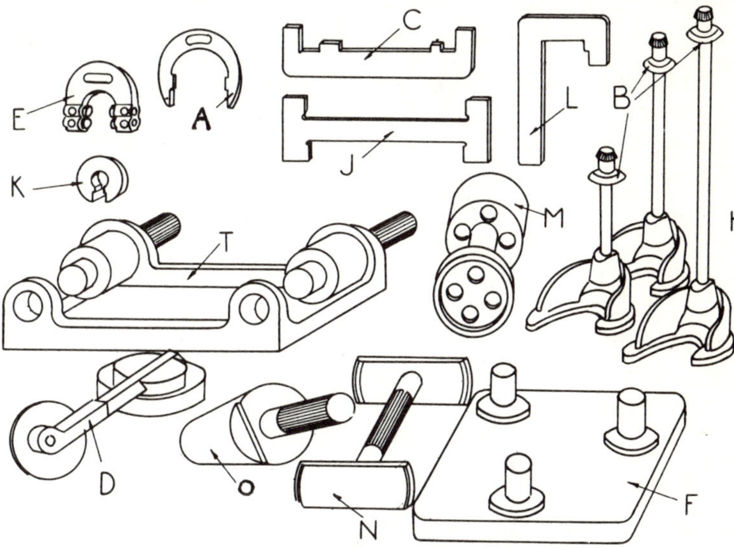

Fig. 11.12 An Assortment of Gauges used in different applications

A is a gap gauge either made from a steel forging or cut from steel plate capable of being later hardened. Alternatively, the gauge can be heavily chromium-plated to give a very hard wear-resisting surface—a method which can also be used to "build up" gauges that have become worn.

E has already been mentioned as an adjustable gap gauge. The pegs or anvils can be screwed in or out to provide the desired limits, and are then locked by means of side-screws. This type of gauge frequently has a thread form cut in the anvils so that it can be used for checking the effective diameters of threads.

T shows a position gauge having two sliding plugs for checking the bores of connecting rods.

At **M** and **N** are illustrated two designs of gauge for checking bores. **N** would be better used if the bore is very shallow, but it would not check for roundness. It has the advantage of being very light, its stem being a piece of steel tube.

The gauge pictured at **O** is used for checking a tapered bore, the step ground into its upper face providing the required limits. Its knurled handle, which is made separate from the body, can alternatively be made tubular.

All gauges, whatever the purpose for which they are to be used, should be made as light as possible. This helps the operator to get a more accurate "feel" of the work, and is also less tiring.

11.12 Materials. Since gauges are generally expensive tools (the greater the degree of accuracy required, the greater normally being their cost), the material of which they are made should be such that the danger of wear—and of resulting inaccuracy—is reduced to a minimum. Tool or carbon steel gives the best hardness and anti-wear properties, but there is always the risk that it may crack in the hardening process and so ruin an expensive gauge. This danger can be partly overcome by using mild or case-hardening steel, and then hardening to instructions supplied by the makers.

In the case of plate gauges, of course, only those parts on which wear is liable to take place need to be hardened at all. For thread gauges, steel should be selected that will harden at a low heat, to reduce the risk of the threads warping or otherwise becoming distorted.

Although gauges of above-average size can be usefully lightened by making parts of them in aluminium, this should be only done with caution lest the correct relationship of the more important parts of the gauge be thrown out of true by an accidental blow or strain. Aluminium is a relatively soft metal, and cannot be appreciably hardened by heat treatment.

11.13 Gauge Accuracy. For National Standard or laboratory work, flat-ended length standards can be produced to an accuracy of one part in a million, but such accuracy as this is fortunately not required for everyday production. Reference bars and standard end-measuring blocks or slip gauges can be made accurate to within 2·5 millionths of their own length; but for general shop use, bars and gauges accurate to within 5 millionths of their length are normally adequate for the purpose and can be bought at lower cost.

Cylindrical gauges up to 25 mm in diameter can be made to within $\pm 0{\cdot}0006$ mm, but such accuracy as this is not normally required. A generally acceptable level of accuracy would be around $\pm 0{\cdot}001$ mm for a lapped gauge. Larger sizes, of course, will be proportionately less accurate in their actual dimensions.

Snap gauges (the fixed type of external-limit gauge) cannot be made closer than ±0·0025 mm because of the "whip" inherent in their structures; and gauges for taper holes and shafts are not likely to be produced nearer than ±0·0025 mm at any diameter along the length. Position or "receiver" gauges having perhaps several points of location and more than one position on the work to be controlled are not likely to be accurate to better than ±0·025 mm of the required dimensions, and any greater accuracy required would involve considerable cost.

When greater than normal accuracy is required, it is the practice to check the work with the aid of special instruments called *comparators*, of which several patented arrangements exist. In principle, the comparator measures the small difference between the size of the work and that of a standard end-measuring block or slip gauge nearly equal to it in size, and then magnifies the difference so that it can be easily corrected.

Details and illustrations of available comparators are given from time to time in the technical press. They fall outside the comparatively limited scope of this book.

11.14 Allowances for Wear. Allowances for wear must always be kept as small as possible. It is true that the larger the allowance, the longer can the gauge wear without producing scrap work. But it is also true that until the gauge has worn considerably, work will have to be produced to a smaller tolerance than is necessary for adequate functioning—and the cost of such work will be higher than it need be because of the replacement and re-setting of the sizing tools that will be required. When the work tolerance is as large as (say) 0·10 mm, it would be safe to design in 0·005 mm for wear; but for smaller tolerances 0·0025 mm is as much as should be allowed.

It is generally true that the harder the gauging surface, the longer time will elapse before wear is apparent (or, alternatively, the less will be the amount that has to be filched from the work tolerance for gauge wear). To obtain extra wear, many gauges are plated with hard chromium, which gives them from two to four times the "life" of the more usual carbon-steel gauge, hardened and lapped.

It has been estimated that standard slip gauges wear approximately 0·00025 mm for every 1,000 times they are wrung together. The high quality of their surface finish causes them to hold together by means of a form of physical adhesion between their surfaces which in its effects resembles magnetic attraction.

CHAPTER 12

The Organisation and Economics of Production

12.1 Earlier chapters in this book have discussed the design and applications of jigs, fixtures and tools, and the relationship of each to the scale of production required and to the type of machine tool available for use. In Chapter 1, job, batch and continuous production requirements were compared, and it was said that the decision on how much (if any) special tooling should be designed for a particular job should be based entirely on cost—*i.e.*, with the sole object of ensuring that unit manufacturing cost is kept to a minimum.

It is the purpose of this final chapter to review the position which the Jig and Tool Department occupies in the manufacturing process, and its responsibilities to other departments in the organisation. Finally, the way in which factory costs are built up will be considered, with particular reference to the effects on component cost of the quantity and quality of production.

12.2 Management Organisation. Fig. 12.1 is a typical organisational chart for a large manufacturing company, in which the detailed responsibilities of the Production Manager are alone of current concern. It will be seen that these responsibilities fall into two groups —Production Planning, as detailed on the left half of Fig. 12.1, and Production Control on the right half. Note that Jig and Tool Design is a Planning activity which has to operate *ahead* of production in order to ensure that no hold-up occurs.

In a smaller works, several of the functions for which the Production Manager is responsible may be undertaken by a single subordinate—for instance, work study, process planning and estimating. These functions are to be considered as "services" to the manufacturing departments. Their job is to supply information on method of manufacture, a schedule of tools and materials required, and a record of progress and output.

12.3 Functions of Jig and Tool Department. When a new product is to be put into manufacture, the Process Planner will decide the method of manufacture in relation to the size of the order, the rate of output necessary and the quality of finish required; and he will

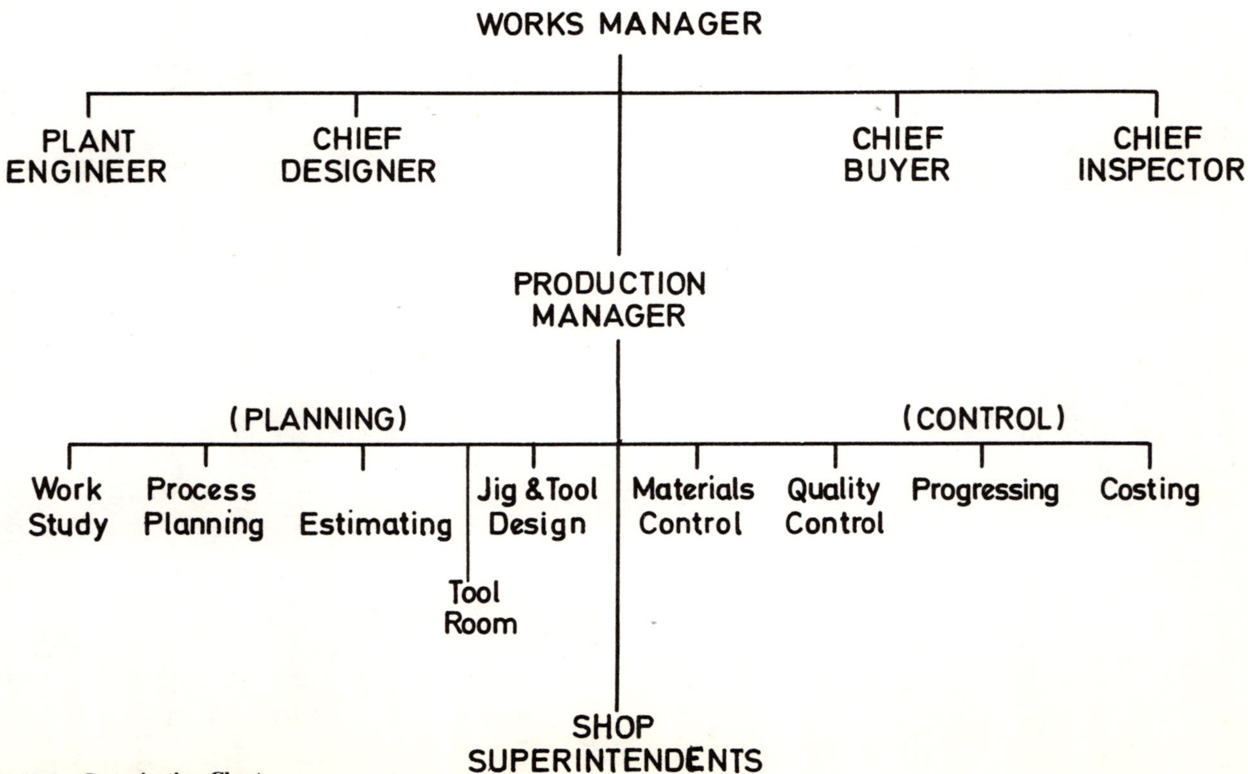

Fig. 12.1 Organisation Chart

make his decision on the basis of minimum cost. After liaison between all the Planning groups, a decision will be made whether to use for the job machinery already available in the works, or whether new machines will have to be bought. In either case, work will fall on the Jig and Tool Department to design, and then to make, any special tooling required by the machines to be used; and there will often be pressure from the manufacturing departments to get this work done as quickly as possible so that production can begin at the earliest moment.

From time to time suggestions will also reach the Department from the Work Study Engineer concerning planned changes in methods of manufacture—be it for continuous production or for repeated batch work. These suggestions, too, will put a load on the Jig and Tool Department; but in this case it is unlikely that the time-scale will be so urgent.

12.4 Manufacturing Costs. In manufacturing any article for sale, it is necessary to allocate accurately all the costs of production associated with it in order to determine a selling price at which an acceptable profit can be made. This is the job of the Estimator. It demands not only a good engineering training, but also detailed knowledge of overhead costs both of the factory itself and of supporting services such as the Design, Accounts and Sales Departments.

The materials and labour which actually go into doing a job are called *direct* costs, and together they constitute **prime cost.** Once the manufacturing processes to be used and the acceptable quality level have been fixed, this prime cost will be the same for every article, and will be directly proportional to the number produced.

The next costs to be considered are those for jigs and tools, heat, light and power, and other consumable materials such as oil, rags, etc. Collectively, these are called *indirect materials. Indirect labour* comprises supervision, inspection and clerical back-up, as well as internal transport, maintenance and stores. It will be seen that some elements of these **indirect costs** will vary with the quantity of the manufactured article to be made, but others will not. A jig which may be good for making 5,000 components, for instance, will need to have its cost written off over only 1,000 if that is the extent of the full order. Nor is it generally possible to adjust the cost of supervision in line with changes in output.

Indirect costs can therefore be called semi-variable, having a fixed minimum irrespective of quantity of output, plus a variable factor which for simplicity is usually assumed to be proportional to output.

Lastly, there are the **fixed costs** of running the factory—made up of rent, rates, depreciation of buildings and machinery, etc., plus the cost of staffing the so-called "non-productive" departments such as design, accounts and personnel. Other costs to be added include sales, publicity and general administrative expenditure. Whereas the indirect costs of manufacture can be allocated fairly accurately to individual items of manufacture, this is much more difficult to do with the fixed costs.

Various methods are used to achieve a rational distribution of these overhead costs, two of the most common being to add them to the direct labour cost (which gives a man-hour rate for productive labour), or to add them to the machine costs (which give a machine-hour rate for each machine or groups of machines).

All costs other than prime cost (*i.e.* indirect costs plus fixed costs) are collectively called **overheads** (sometimes alternatively referred to as **oncost** or, more rarely, **burden**). It follows therefore that, for every article manufactured,

Prime Cost + Overheads + Profit Desired = Selling Price.

It is not difficult to see that, starting from a given Selling Price, Profit will tend to rise with output; and the factors which govern this correlation will be examined in the next section.

12.5 Break-Even Chart. The object of the chart, Fig. 12.2, is to show the balance of costs against sales over a fixed period of time (usually a year or less). Against the planned output are set the fixed overheads, the variable overheads, and the prime cost. After allowance is made for the fall of unit cost as output rises, a "total cost" line is obtained; and break-even point is established at the point where the sales line crosses this total cost line.

It becomes apparent that quite small changes in the level of output (if the extra output can be sold without price concessions) will have a disproportionately large effect on the margin of profit—indeed, a 20 per cent increase in output, if it can be sold, will often double the overall profit. This fact, as revealed by the chart, is of great help to the management in assessing, *e.g.*, the advantages of overtime and double-shift working. It will also become apparent how quickly money spent on improved tooling will repay itself if it results in a reduction in manufacturing or assembly time.

12.6 Unit Cost. The cost of each unit of production varies with the rate of output, with the number of units produced, and with the level of quality standards demanded.

(1) *Rate of Output*. Because some overheads are fixed and so do not vary with output, unit production costs must fall with increased productivity. The Jig and Tool Department, working in close liaison with the Work Study section, can play a very important role in bringing down costs in this way.

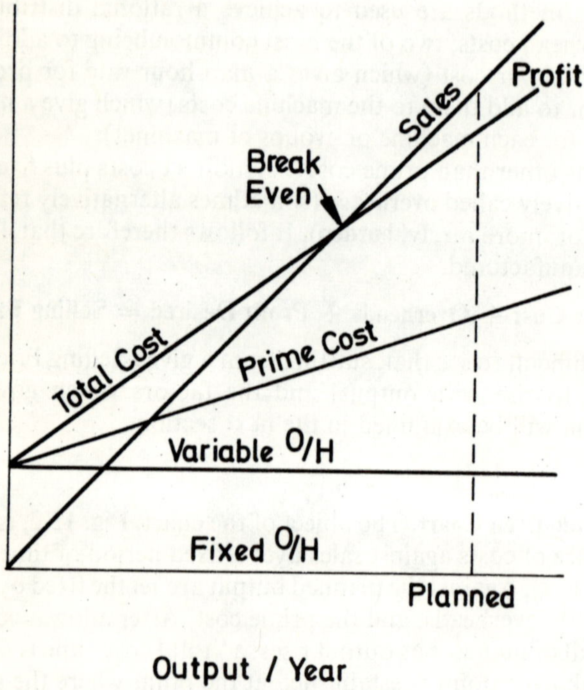

Fig. 12.2 Break-even Chart

(2) *Size of Order*. In addition to the actual rate of output, the total quantity on order often has a marked effect on unit costs. The larger the order, the greater can be the investment in capital equipment and special tooling. Take the example of making a component by turning down from bar stock. For a small order a centre lathe would be used; larger numbers would justify setting up a capstan lathe; while for mass production the purchase of an automatic lathe would be economically desirable. This is a point which should always be

brought home to the customer, who may well be prepared to place firm repeat orders if a lower price per unit can be offered him.

(3) *Quality Standards.* Falling under this heading are tolerance levels, surface finish and material specifications. The designer should choose the cheapest material which will meet the requirements of strength, rigidity, resistance to corrosion, etc., laid down by the customer—bearing in mind that it is not only the cost of the material itself which is important but also the cost of processing it (high-strength steels are often difficult to machine, and so on).

Tighter tolerances and better surface finish inevitably increase cost, either because it becomes necessary to use more expensive equipment—a jig borer, say, in place of a standard drilling machine—or to employ more highly skilled labour, or because additional finishing processes such as grinding or lapping have to be applied. Fig. 12.3

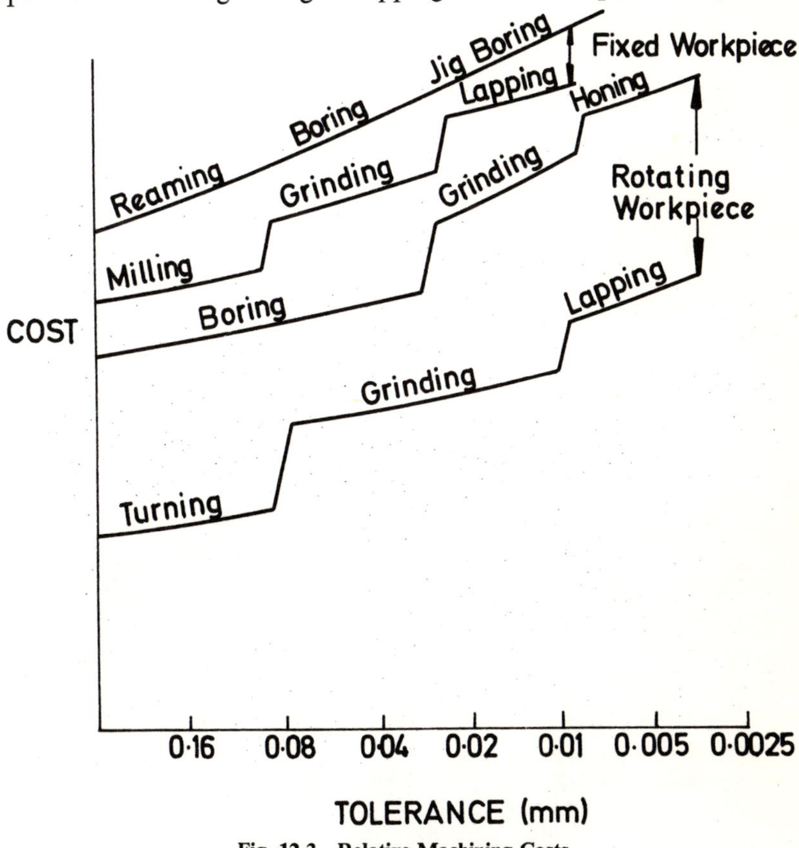

Fig. 12.3 Relative Machining Costs

shows in diagram form how the cost of the different methods of finishing a component vary with the quality of surface finish stipulated by the customer. Though inevitably approximate and liable to vary with circumstances, the orders of magnitude given may well constitute one of the most practically valuable illustrations in this book.

It lies within the power of the Jig and Tool Department to make a major contribution to lower costs if it can contrive that a high standard of accuracy and finish is consistently produced by semi-skilled labour using only a standard range of machine tools.

Index